Proceedings in Life Sciences

Evolution and Genetics of Life Histories

Edited by
Hugh Dingle and Joseph P. Hegmann

With Contributions by

J. ANTONOVICS, W.S. BLAU, B.P. BRADLEY,
C.K. BROWN, R.W. DOYLE, J.T. GIESEL,
J.P. HEGMANN, C.A. ISTOCK, C.E. KING,
R. LANDE, M. MANLOVE, T.R. MEAGHER,
P. MURPHY, R.A. MYERS, R.J. SCHULTZ,
C.A. TAUBER, M.J. TAUBER, A.R. TEMPLETON

With 47 Figures

Springer-Verlag
New York Heidelberg Berlin

Hugh Dingle
Program in Evolutionary
    Ecology and Behavior
Department of Zoology
University of Iowa
Iowa City, Iowa 52242 U.S.A.

Joseph P. Hegmann
Program in Evolutionary
    Ecology and Behavior
Department of Zoology
University of Iowa
Iowa City, Iowa 52242 U.S.A.

Production: Marie Donovan

On the front cover: The figure on the front cover is a three-dimensional projection of the genetic correlations among life history variables of the milkweed bug, *Oncopeltus fasciatus*, sampled from Iowa. (See Chapter 10 for discussion.)

Library of Congress Cataloging in Publication Data
Main entry under title:
Evolution and genetics of life histories.
    (Proceedings in life sciences)
    "This volume is the result of a Symposium entitled "Variation in Life Histories: Genetics and Evolutionary Processes" sponsored by the Program in Evolutionary Ecology and Behavior of the University of Iowa and held in Iowa City on October 13 and 14, 1980"—Pref.
    Bibliography: p.
    Includes index.
    1. Evolution—Congresses. 2. Variation (Biology)—Congresses. 3. Genetics—Congresses. I. Dingle, Hugh, 1946-   . II. Hegmann, Joseph P. III. Program in Evolutionary Ecology and Behavior (University of Iowa). IV. Title. V. Series.
QH359.E9   575   82-3188
                    AACR2

Printed in the United States of America

9 8 7 6 5 4 3 2 1

ISBN 0-387-90702-5   Springer-Verlag New York Heidelberg Berlin
ISBN 3-540-90702-5   Springer-Verlag Berlin Heidelberg New York

Preface

This volume is the result of a symposium entitled "Variation in Life Histories: Genetics and Evolutionary Processes" sponsored by the Program in Evolutionary Ecology and Behavior of the University of Iowa and held in Iowa City on October 13 and 14, 1980. Prompted by a recent upsurge of interest in the evolution of life histories, we chose this topic because of the obvious association between life history traits and Darwinian fitness. If such an association were to be fruitfully investigated, it would require the closer cooperation of population and evolutionary ecologists and quantitative and population geneticists. To encourage such an association, our symposium had four major aims: first, to facilitate intellectual exchange across disciplines among an array of biologists studying life histories; second, to encourage exploration of genetic variance and covariance for life history traits; third, to consider the ecological background for genetic variability; and finally, to facilitate a comparative overview both within and among species.

Obviously such broad aims cannot be met totally in a single volume, but we think we have succeeded reasonably well in providing a representative and nourishing intellectual feast. We see this book as a stimulus to the coordination of future efforts in an important and expanding area of inquiry. We have divided the book into six sections. The first opens with two chapters on the theoretical background provided by population biology and quantitative genetics, and is followed by two sections in which the roles of physiological adaptation and modes of reproduction, respectively, in life history evolution are considered. These are in turn followed by sections examining the quantitative genetics of life histories, assessing variation first within populations and then among populations and species. The final section is an edited transcription of the concluding discussion, which we hope captures the flavor of the symposium and the stimuli and challenges to future research that resulted. In any event, we hope readers of this book learn from its chapters as we have and are encouraged to tackle some of the difficult but fascinating problems presented.

The chapters of this volume are expanded and, in some cases, revised versions of all but one of the papers delivered in Iowa City. Regrettably, Dr. Patricia Werner felt it necessary, because of time constraints, to withdraw her paper from this volume. All papers were reviewed by referees chosen by the editors, and we feel that the quality of this book has been enhanced substantially as a result. For their thorough and perceptive reviews of manuscripts, we wish to thank Roger Milkman, David Hamill, James F. Leslie, Michael Rose, Michael Wade, John M. Emlen, Roger Doyle, Conrad Istock, Michael Lynch, Egbert Leigh, Stevan Arnold, Alan Templeton, Janis Antonovics, John Endler, J. David Allan, and Charles Oliver. We owe a special debt of thanks to Stevan Arnold, Michael Lynch, and Barbara Schaal, who accepted on short notice and without hesitation the challenging task of leading the final discussion.

We also wish to thank the many people who made the task of organizing and editing so much easier for us. We owe special debts to our families and our students, who often bore the brunt of inconvenient absences for planning and editing sessions, and who were supportive and reassuring in the midst of attacks of panic. Bob Davis and the staff of the University of Iowa Center for Conferences and Institutes professionally and skillfully handled the logistics of the symposium. Deena Staub, Robin Wagner, and Carol Wolvington helped with editing and typing in Iowa City. Mark Licker and Marie Donovan of Springer-Verlag guided the book through the various stages of publication with their usual skill. Finally, we should like to thank the authors both for intellectual pleasure and for tolerating our impatience with equanimity.

Iowa City   Hugh Dingle
March 1982   Joseph P. Hegmann

Contents

List of Contributors

*STEVAN J. ARNOLD   Department of Biology, University of Chicago, Chicago, Illinois 60637, U.S.A.

JANIS ANTONOVICS   Department of Botany, Duke University, Durham, North Carolina 27706, U.S.A.

WILLIAM S. BLAU   Program in Evolutionary Ecology and Behavior, Department of Zoology, University of Iowa, Iowa City, Iowa 52242, U.S.A.

BRIAN P. BRADLEY   Department of Biological Sciences, University of Maryland, Baltimore, Maryland 21228, U.S.A.

CARL KICE BROWN   Program in Evolutionary Ecology and Behavior, Department of Zoology, University of Iowa, Iowa City, Iowa 52242, U.S.A.

HUGH DINGLE   Program in Evolutionary Ecology and Behavior, Department of Zoology, University of Iowa, Iowa City, Iowa 52242, U.S.A.

ROGER W. DOYLE   Department of Biology, Dalhousie University, Halifax, Nova Scotia B3H 4J1, Canada

JAMES T. GIESEL   Department of Zoology, University of Florida, Gainesville, Florida 32611, U.S.A.

JOSEPH P. HEGMANN   Program in Evolutionary Ecology and Behavior, Department of Zoology, University of Iowa, Iowa City, Iowa 52242, U.S.A.

*Discussion leader

CONRAD A. ISTOCK   Department of Biology, University of Rochester, Rochester, New York 14627, U.S.A.

CHARLES E. KING   Department of Zoology, Oregon State University, Corvallis, Oregon 97331, U.S.A.

RUSSELL LANDE   Department of Biophysics and Theoretical Biology, University of Chicago, Chicago, Illinois 60637, U.S.A.

*MICHAEL LYNCH   Department of Ecology, Ethology, and Evolution, University of Illinois, Urbana, Illinois 61820, U.S.A.

MICHAEL MANLOVE   Department of Zoology, University of Florida, Gainesville, Florida 32611, U.S.A.

THOMAS R. MEAGHER   Department of Botany, Duke University, Durham, North Carolina 27706, U.S.A.

PATRICIA MURPHY   Department of Zoology, University of Florida, Gainesville, Florida 32611, U.S.A.

RANSOM A. MYERS   Department of Biology, Dalhousie University, Halifax, Nova Scotia B3H 4J1, Canada

*BARBARA SCHAAL   Department of Biology, Washington University, St. Louis, Missouri 63130, U.S.A.

R. JACK SCHULTZ   Biological Sciences Group, University of Connecticut, Storrs, Connecticut 06268, U.S.A.

CATHERINE A. TAUBER   Department of Entomology, Cornell University, Ithaca, New York 14853, U.S.A.

MAURICE J. TAUBER   Department of Entomology, Cornell University, Ithaca, New York 14853, U.S.A.

ALAN R. TEMPLETON   Department of Biology, Washington University, St. Louis, Missouri 63130, U.S.A.

*Discussion leader

Introductory Chapter

Genetics, Ecology, and the Evolution of Life Histories

HUGH DINGLE and JOSEPH P. HEGMANN*

In a seminal discussion MacArthur and Wilson (1967) provided a logical framework for the development of a theory of life history evolution. They reasoned that colonizing species at low densities in pioneer habitats should evolve a suite of life table characters emphasizing early and rapid production of offspring. In contrast, species in stable communities with density dependence were expected to display traits associated with strong competitive abilities and behavioral characteristics promoting the survival of individual offspring. Their exposition clearly emphasized the importance of gene influences on life history variation, because an understanding of genetic structure is essential to understanding evolutionary processes. In their words, "The degree of deflection and rebound of r depends of course on the heritability of the life-table parameters. Measurements of this heritability, and estimates of its influence on evolution under various colonizing conditions, remain to be made." Until recently, additional analyses of life history evolution have been largely theoretical and usually without the explicit inclusion of appropriate genetic variation (summarized by Stearns 1976).

In spite of this lack, life histories provide rich material for the unfolding plot of the evolutionary play involving genetics and ecology. In the first place, they are "of unique importance to general Darwinism" (Bell 1980) because fitness is an inevitable and calculable consequence of the schedule of births and deaths, the major phenotype in life history evolution. Second, recent empirical research on life table statistics has shown quasicontinuous variation featuring polygenic systems and fluctuating environments (Dingle et al. 1977, Giesel 1976, Hoffman 1978, Istock 1978). The dynamics of evolving organisms lie in the action of natural selection on varying phenotypes and hence in the ways those phenotypes depend on the structure of the underlying genotypes. These relationships are the proper subject of quantitative genetics, and the major focus of this volume is the attempt to bring together quantitative genetics and ecology for the analysis of a major fitness phenotype, the life history. In this endeavor we differ from other recent attempts to draw genetics and ecology together as these

*Program in Evolutionary Ecology and Behavior, Department of Zoology, University of Iowa, Iowa City, Iowa 52242 U.S.A.

have been directed primarily at population genetics (Berry 1979, Brussard 1978, Hedrick et al. 1978).

There has been increasing commitment to the proposition that it is necessary to determine the importance of gene effects before a complete life history theory can be developed (Istock 1978, Stearns 1976). Recent empirical (Dingle et al. 1977, Istock et al. 1976, McLaren 1976) and theoretical (Lande 1979) studies suggest that quantitative genetics indeed provides the most appropriate starting point for investigating complex adaptations such as life histories. Although theory and methods may not yet be sufficient to deal with the products of evolution in the form of species differences (but see Lande 1980), they do allow analysis with predictive validity of genetic variance and covariance for key fitness variables within species (Dingle et al. 1977, Istock 1978). In this way they may permit fuller development of the evolutionary theory of life histories. The authors in this volume have contributed to our embryonic knowledge of genetic variance and covariance for life history traits and explore the topic further here.

In contrast to genetic studies, there has been extensive development of sophisticated ecological theory with respect to life histories (Bell 1980, Roughgarden 1979, Stearns 1976). Emphasis, however, has been on means and the assumption that under heterogeneous environmental regimes a constant proportion of the population always breeds (Nichols et al. 1976 treat this fully). Our emphasis in this volume is on variation, and the contributors have been chosen because they have addressed problems of variation in an ecological context and have considered evolution with respect to important environmental heterogeneities. The ecological background to genetic variability is an important element of this book.

A diversity of organisms has contributed to both genetic and ecological studies of life histories with great potential for comparative analysis. Such analysis requires synthesis and interaction among evolutionary biologists with interests as diverse as the organisms they study. The organisms discussed in this volume come from marine, fresh water, tropical, and temperate habitats. So far as we know, there has been no attempt to compare genetic structure of life histories across species. One of our intents has been to present this kind of comparative overview in the hope that broadened perspectives may facilitate novel insights and refinement of theory.

In summary, we attempted to bring together, in the symposium and as contributors to this book, investigators whose research constitutes a significant proportion of the recent contributions to understanding genetic variation in life history traits. Life history theory is, for the reasons given above, an important topic in evolutionary biology, and current interest in its development seems to us, at least, to make this volume particularly appropriate at this time. We hope these collected papers stimulate expanded research efforts on the complex but important issues they raise.

References

Bell, G.: Costs of reproduction and their consequences. Am. Nat. 116, 45-76 (1980).
Berry, R. J.: Genetical factors in animal population dynamics. In: Population Dynamics. Anderson, R. M., Turner, B. F., Taylor, L. R. (eds.). Oxford: Blackwell, 1979, pp. 53-80.

Brussard, P. F. (ed.): Ecological Genetics: The Interface. New York: Springer-Verlag, 1978.

Dingle, H., Brown, C. K., Hegmann, J. P.: The nature of genetic variance influencing photoperiodic diapause in a migrant insect, *Oncopeltus fasciatus*. Am. Nat. 111, 1047-1059 (1977).

Giesel, J. T.: Reproductive strategies as adaptations to life in temporally heterogeneous environments. Ann. Rev. Ecol. Syst. 7, 57-79 (1976).

Hedrick, P., Jain, S., Holden, L.: Multilocus systems in evolution. In: Evolutionary Biology, vol. II. Hecht, M. K., Steere, W. C., Wallace, B. (eds.). New York and London: Plenum Press, 1978, pp. 101-184.

Hoffman, R. J.: Environmental uncertainty and evolution of physiological adaptation in *Colias* butterflies. Am. Nat. 112, 999-1015 (1978).

Istock, C. A.: Fitness variation in a natural population. In: Evolution of Insect Migration and Diapause. Dingle, H. (ed.). New York: Springer-Verlag, 1978, pp. 171-190.

Istock, C. A., Zisfein, J., Vavra, K. J.: Ecology and evolution of the pitcher plant mosquito. 2. The substructure of fitness. Evolution 30, 548-557 (1976).

Lande, R.: Quantitative genetic analysis of multivariate allometry. Evolution 33, 402-416 (1979).

Lande, R.: Genetic variation and phenotypic evolution during allopatric speciation. Am. Nat. 116, 463-479 (1980).

MacArthur, R. H., Wilson, E. O.: The Theory of Island Biogeography. Princeton: Princeton University Press, 1967.

McLaren, I. A.: Inheritance of demographic and production parameters in the marine copepod, *Eurytemora herdmani*. Biol. Bull. 151, 200-213 (1976).

Nichols, J. D., Conley, W., Batt, B., Tipton, A. R.: Temporally dynamic reproductive strategies and the concept of r- and K-selection. Am. Nat. 110, 995-1005 (1976).

Roughgarden, J.: Theory of Population Genetics and Evolutionary Ecology. New York: MacMillan, 1979.

Stearns, S. C.: Life history tactics: A review of the ideas. Q. Rev. Biol. 51, 3-47 (1976).

PART ONE

Theory

Developing the theory for polygenically based life history evolution requires melding the classical population biology of life tables and reproductive schedules with the methods and concepts of population and quantitative genetics. This can be approached in two ways, one via population biology itself and the other by starting with appropriate genetic models. Whichever approach is taken, it is probably not sufficient to restrict attention to single life history traits. In the first approach, such restriction ignores the true tangled intricacies of population dynamics; in the second it ignores the critical fact of genetic covariance structure. Both contributors to this section deal directly with this issue of multivariate complexity.

In the first chapter Conrad Istock takes the population biology approach and briefly reviews its history. Then, starting with the premise that both biotic and abiotic factors potentially limit populations, he considers the genetic restructuring of life history characters that eventually contribute to irreversible evolutionary change (meso-evolution). From these considerations he develops the notion of "regulative value," the age-specific loss in potential net reproduction resulting from the action of natural selection. Because regulative value derives from genetic and environmental interactions among the entire suite of life history traits, significant changes in life history pattern often may be extensive and complex enough to result in irreversible genetic change and thus mesoevolution.

Russell Lande extends multivariate quantitative genetic theory to incorporate demographic structure for populations with overlapping generations. To do this, he uses a standard matrix approach to population modeling but treats mortality and fecundity as quantitatively varying characters in which evolutionary change depends on response to selection acting on polygenic systems. The change in individual life history characters under selection depends not only on the selection gradient but also on the additive genetic variance-covariance structure relating all life history traits. This means that no life history theory is complete without reconsideration of these multivariate genetic constraints.

Chapter 1

Some Theoretical Considerations Concerning Life History Evolution

CONRAD A. ISTOCK*

Introduction

During the last 10 years evolutionary biology has entered a period of expanding theoretical scope following about two decades of rather settled satisfaction with the "modern" or "neo-Darwinian" synthesis. The modern synthesis is characterized, perhaps above all else, both by the elaboration of mathematical structures to describe the genetic diversity and response to natural selection in populations and by a qualitative focus on the properties of species and the process of speciation. The mathematical models of microevolution explored prior to 1970 usually incorporate relatively simple Mendelian genetic descriptions and ecologically simple selection regimes. The focus on species in the modern synthesis brought a much clearer conception of the biological nature of species and some easily grasped, heuristic, models of allopatric and sympatric speciation.

The past 10 or 15 years have seen the extension or initiation of theoretical explorations into such complex phenomena as the evolution of polygenically inherited traits, the evolution of genetic regulation and development in eukaryotes, the roles of sexuality and genetic recombination in evolution, chromosomal evolution, evolution in fluctuating environments, interdemic selection, geographical differentiation of species populations, coevolution of species, phenotypic evolution, the evolution of many kinds of behavioral patterns, group selection and the evolution of sociality, rates of species origination and extinction, as well as the evolution of life history phenomena. At present, evolutionary theory lacks the degree of coherence that seemed to prevail under the modern synthesis, but in exchange it offers a richness of questions and potential experimentation to be explored for decades to come.

The absence of any direct contact between mathematical representations of microevolution and the processes of speciation and species divergence was tolerated in the formulation of the modern synthesis. Now, a critical issue is whether any quantitative model can represent degrees of evolutionary divergence, at least in a probabilistic form, all the way from the simplest microevolutionary modifications to the formation of dis-

*Department of Biology, University of Rochester, Rochester, New York 14627 U.S.A.

tinct species. Under the modern synthesis there perhaps always has been the fond hope that extensions of the models of classical population genetics would eventually clarify the quantitative relation of selection and speciation. Alternatively, species formation might be viewed as too irregular, infrequent, or heterogeneous a process to be predicted, explained, or even represented by extensions of equations describing directional, microevolutionary change in gene frequencies. This lack of connection between levels and kinds of discourse within the modern synthesis—this blending of a carefully detailed logic for microevolution and a largely intuitive theory for larger scale evolution and speciation—was bound to engender a period of intense reexamination and extension such as we have now entered. Central to these issues is the recognition that sequentially compounded transformations of population genetic structure, whether in one phyletically descendent line or in two or more genetically diverging lines, recurrently achieves irreversibility in a probabilistic sense (Dobzhansky 1970). Irreversibility of genetic change frequently may be enforced by speciation, but the importance of irreversibility derives from the possibility that it will recur in phyletically descendent lines without speciation.

At the moment it seems unwise to break, as Gould (1980) has done, with the intuitive conclusion that irreversible evolutionary changes, near the level of either chronospecies (Stanley 1979) or phyletically contemporaneous species, are nothing more than the accrued weight either of many allelic substitutions of small effect at many loci or of the more pervasive effects of a smaller number of "highly influential" genetic alterations (major gene, regulatory gene, chromosomal rearrangement). The next step is to develop and test mathematical models of evolution that span the hiatus between reversible and irreversible genetic change. I think realistic models of life history evolution fall squarely into this conceptual hiatus. I am suggesting that the sequential genetic changes underlying statistically significant shifts in patterns of survival, longevity, developmental rate, age-specific fertility, semel- vs. iteroparity, seasonality, migration, diapause or dormancy, degree of outbreeding, etc., will typically involve large enough changes of genetic architecture to prohibit a straightforward reversal of accumulated genetic change. Of course, convergence back to a resemblance of an earlier life history pattern may be achieved by a quite new path of genetic reorganization. When only a single life history character undergoes modification, as is possible with artificial selection, the underlying genetic change may be simple and reversible. We have examined such a case by experimentally altering the habitat (oviposition) choice of the pitcher-plant mosquito *Wyeomyia smithii* (Istock et al. 1982). However, when many life history features simultaneously undergo significant remolding, the process is likely to be genetically extensive and quickly achieve irreversibility. It is important for many questions about evolution to know whether this is so. Although at present it is not completely known when reversible and irreversible life history changes occur, the problem may be experimentally approachable with selection experiments. The question of when and how evolutionary change becomes irreversible seems to lie at the heart of the overall question of continuity between microevolution and larger scale evolutionary divergence.

Realism in both the genetic and ecological senses seems to be a prerequisite for improved evolutionary models—models that might make closer approach to empirical information and models of a larger scope, allowing for irreversible genetic change. The microevolutionary models of the modern synthesis were often unrealistic on both counts but developed toward greater and greater detail and realism on the genetic side, although almost always retaining explicitly Mendelizing genetic determinants. The same models were always severely limited in ecological detail; treating selection as a disembodied force, rather than as the consequence of explicit ecological conditions and attendant demographic patterns. Rare exceptions such as Levene's (1953) multiple niche model for the maintenance of genetic polymorphism, or the Haldane and Jayakar (1963) model of temporally fluctuating selection, achieved striking results with only tiny ventures toward ecological realism.

During the first 20 years under the "modern synthesis," ecological models of life history evolution contained no genetic elements at all. Early models, particularly the pioneering work of Cole (1954), involved extensive demographic detail but then backed away from the difficulties thus imposed through unrealistic simplifying assumptions such as perfect survival. Still, important insights were found, in particular the focus on the sharp dichotomy between semelparity and iteroparity so often seen in nature. Even in recent years the dominant tendency of life history theorists has been to move to extreme simplification on the ecological side and little or no addition of genetic representation. The introduction of the concept of r vs. K selection brought some order to the way evolutionary ecologists tallied the life history attributes of species, but for reasons of logic (Stearns 1977) the r vs. K dichotomy seems trapped away from stimulating the development of realistic and mathematically complete models. An instructive exploration of the conditions for the evolution of semelparity and iteroparity was achieved by approaches simpler than Cole's (Bell 1976, Murphy 1968, Schaffer 1974), and these efforts pointed out the need to consider fluctuating environments, reasoned in interesting ways about optimality of life history patterns (but see Lewontin 1979), but did not attempt inclusion of genetic descriptions, as lamented by Stearns (1976).

In an earlier publication I tried to combine a demographic representation of the ecological side of natural selection with a polygenic inheritance of life history characters (Istock 1970). The demographic approach was an attempt to put greater ecological realism into Cole's (1954) approach, and the choice of a polygenic representation of inheritance was influenced by Fisher's (1958) theoretical development of the genetic basis for the fundamental theorem of natural selection. In parts of this chapter I want to revise and extend this earlier model involving "regulative value." Obviously, I do this because I hope this line of mathematical reasoning will someday help us across the hiatus between microevolution of the single-gene substitution sort and the lowest level of irreversible genetic change under selection, or what Dobzhansky (1970) called mesoevolution. At this stage the model falls short of that vaunted goal, but it does serve to identify some of the intriguing problems we face.

Complexity in the Suite of Life History Characters

Life history and fitness characters are synonymous insofar as they exert direct influence over expressed reproduction. Unfortunately, this obvious connection between ecological and evolutionary theory does not immediately supply much clue to the process by which life history patterns evolve because it is not clear how fitness is best defined. At various times mathematical ecologists have used the intrinsic rate of increase, reproductive value, net reproductive rate, population size, and related measures based on mathematical demography as the measure of fitness. Recently, Tuljapurkar and Orzack (1980) have questioned whether a single realistic measure of fitness is possible and have argued that extinction probability may be an essential consideration along with other more traditional measures. Thoday (1958) much earlier proposed a probability of long-term extinction as a single measure of fitness and it is possible to incorporate his "existential" (Istock 1981a) definition of fitness in either theoretical (Istock 1981b) or empirical, laboratory (Saul 1970) studies of microevolution. It remains to be seen how the notion of extinction probability can be brought into models of irreversible evolutionary change.

The prime life history or fitness characters are survival, fertility, reproductive span, degree of iteroparity, seasonality, and extent of sexuality. The last character exerts much influence over the role of genetic variation and genetic recombination in the course of evolutionary change. All but the last of these characters can be at least partly described and measured demographically. Although the complexity of a comprehensive life history description is substantial and as yet beyond the reach of theoretical endeavors, such complexity is not likely to be overwhelming. An additional, particularly interesting, and frequently recognized aspect of this complexity is the phenotypic and genetic covariance that may exist among different life history characters. It remains unclear how much the number of dimensions ascribable to life history phenomena and their determining genetic structures can be reduced while still achieving more successful evolutionary models. Future models with increased realism will almost certainly give way in the direction of much greater dimensionality. As a minimum, two character sets, the life table and the fertility table (vital statistics), together can encompass a great deal of the information in any life history pattern. On the genetic side, polygenic measures such as additive variance, heritability, and genetic correlation or covariance may summarize some of the complexities to be found among the genetic determinants and developmental processes that underlie life history phenomena.

Polygenic Inheritance of Life History Characters

Evidence from a wide variety of laboratory studies and a few field studies suggest that polygenic variation for life history or fitness characters is widespread and moderately heritable in populations. However, it is unclear how much of this variation is continually exposed in natural populations and how much of this variation is exposed on environmental change of the magnitude experienced when stocks from nature are transferred to laboratory experiments (Istock 1982). It is also of interest that this polygenic variation is not uniformly present from one life history character to another, or even within one type of character over different ages. For example, the amounts of genetic variation for development time in different parts of insect life cycles are far from uniform (for details of the above phenomena see Istock 1982). In some fortunate cases it may be found that life history features are controlled by simple Mendelian inheritance at one or two loci (Tauber and Tauber 1978 and Chapter 4, this volume), but for the majority of cases polygenic inheritance appears to be the rule.

A Demographic Representation of Age-Specific Selection

The Ecological Aspect

I next want to characterize the ecological side of natural selection using a demographic representation of the way in which selection may scan the expressed life history of a population in any generation. The model revolves around the potential value of each age for increasing the expressed, average value of the net reproductive rate. It accomplishes this age by age "valuation" by juxtaposing the greatest possible survival and fertility in the prevailing environment with average, expressed survival and fertility, and it weights the departures among these two sets of vital statistics according to the implicit loss in potential net reproduction. At each age the loss to maximum reproduction is called the "regulative value" of that age. The model considerably reduces the dimensionality of life history phenomena by concentrating on only age-specific survival and fertility as descriptors of a life history.

Fisher's (1958) "reproductive value" was the first attempt to solve the problem of average, age-specific, valuation of individuals in a population under natural selection. As a measure of the age-specific effect of selection, reproductive value is largely prospective. It takes as given and fixed the current average survival and fertility of a population and depicts, ceteris paribus, the relative contribution of each age to future population size. As such, reproductive value does not capture the "here and now" aspect of selection or the "room for improvement," so to speak, latent in the prevailing ecological circumstance and the continuing birth and death events that can accomplish the evolutionary transformation of a life history pattern.

In contrast, regulative value is meant to assess the selective consequences of the birth and death events during each biological generation. As will be seen, the method is both retrospective and prospective. Another way to render the meaning of regulative value is to consider it an age-specific apportionment of unrealized, but potential, net reproductive rate, or $R_{0,max}$. This apportionment is done as follows.

Consider a population in a consistent environment, E_1. A consistent environment, unlike a constant one, may exhibit spatial or temporal heterogeneity, but all ecologically relevant variables, other than population size, have level averages over time. This conception of environment allows for the possibility that polygenic variation (complex genetic polymorphisms) may persist via forms of stabilizing selection but ensures that the vital statistics from which are gauged the directional path of life history evolution are fixed. Imagine now that the maximum values for survival and fertility are obtained under the physical-chemical conditions of E_1 and are attributable to one or more of the phenotypes in the population. These maximal values are the result when all biotic restrictions from predators, pathogens, competitors, parasites, or shortages of food, nutrients, and shelter are absent. Further, these maximal values are appropriate for a generation in which the average expressed survival and fertility is observed at each age when all prevailing ecological restraints influence the population. For females, therefore, are defined:

L_x: Maximal age-specific survival, potentially achievable by one or more female phenotypes aged x and living under the physical chemical conditions of E_1, but with all biotic restraints absent;

M_x: Maximal age-specific fertility potential for one or more phenotypes, all biotic restraints absent;

ℓ_x: The average, natural, or realized age-specific survival for the population expressed in the current generation, all biotic restraints included;

m_x: The average, realized, age-specific fertility expressed in the current generation, all biotic restraints included;

ℓ'_x: The missing (imaginary) average, age-specific survival from unrealized, but phenotypically feasible, survival rates;

m'_x: The missing (imaginary), average fertility from unrealized, but phenotypically feasible, births to females aged x.

L_x and M_x set the upper, physiological bounds to survival and fertility, and ℓ_x and m_x represent the realized life table and fertility table observed in nature. E_1 characterizes the physical and chemical conditions of environment for both sets of demographic values. Therefore, potential and actual performances of a population are defined. Their relation to each other is illustrated in Figure 1-1.

Now the problem is to determine the deduction from $R_{0,max}$ occurring at each age, i.e., the amount of restraint or regulation falling at each age i. In the notation to follow, x serves as the index of age for successive entries in any representation of vital statistics, j is the index of age for a cumulative product, and i denotes a specific age.

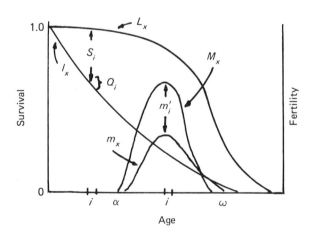

Figure 1-1. Graphical representation of the various age-specific, survival, and fertility functions used in the regulative value model. An arbitrary prereproductive age i, with its associated values of $S_i = L_i - \ell_i$ and Q_i, is shown toward the left end of the age axis. An arbitrary reproductive age i with associated $m'_i = M_i - m_i$ is shown in the middle of the age axis. Any reproductive age will also have its respective S_i and Q_i values. See text for complete definitions and further explanation.

Reasoning from the definitions and Figure 1-1, the total difference in survival at any age i is:

$$S_i = L_i - \ell_i$$

of which part may result from mortality increments, Q_x, at previous ages so that the loss from deaths solely at age i is:

$$Q_i = S_i - \sum_{x=0}^{i-1} Q_x = \ell'_{ii}$$

The imaginary life table ℓ'_x for age i is then constructed using $Q_i = \ell'_{ii}$ as the radix, or first entry, so that for all $x > i$:

$$\ell'_{ix} = \ell'_{ii} \prod_{j=ii}^{x-1} P_j$$

with the P_j as age-specific probabilities along the L_x curve. Each P_j is defined as:

$$P_j = L_{j+1}/L_j$$

Thus, each ℓ'_{ix} column measures the partial departure from performance along the L_x curve resulting solely from deaths at age i. The loss in net reproduction from deaths solely at age i is then estimated by taking the summed cross product of ℓ'_{ix} with M_x.

Let α be the age of first reproduction for the fastest developing phenotype in E_1, i.e., the first $M_x > 0$, and let ω be the greatest reproductive age for any female in E_1, i.e., the last $m_x > 0$ or $M_x > 0$, whichever comes at the greater age. When i is a potential reproductive age there is the further loss of net reproduction from $m'_i = (M_i - m'_i)$. At all other i, $m'_i = 0$. Any losses of potential reproduction through the unrealized births m'_i are associated with the corresponding ℓ_i value, the survivors who might have added m'_i more births at age i, so the additional loss of net reproduction at age i by this process is $\ell_i m'_i$.

The total loss or cost, in units of net reproduction, attributable to birth and death events at age i then becomes:

$$c_i = \ell_i m'_i + \ell'_{ii} M_i + \sum_{x=i+1}^{\omega} (\ell'_{ii} \prod_{j=i}^{x-1} P_j) M_x$$

The first term on the right accounts for unrealized births, the second term accounts for loss of reproduction specifically as a result of deaths at age i, and the third term accumulates all increments of potential reproduction lost over all reproductive ages greater than age i. If the "imaginary" life table ℓ'_{ix}, including ℓ'_{ii} as its first entry, is calculated ahead of time the expression c_i is written more simply as:

$$c_i = \ell_i m_i' + \sum_{x=i}^{\omega} \ell_{ix}' M_x$$

If c_i is called the "regulative amount," and if the total restraint of the population in E_1 is $\sum_{x=0}^{\omega} c_x$, the relative "regulative value" of the ith age is defined as:

$$v_i' = c_i / \sum_{x=0}^{\omega} c_x$$

with the total consequence being:

$$\sum_{x=0}^{\omega} c_x = R_{0,max} - R_{0,nature}$$

where

$$R_{0,max} = \sum_{\alpha}^{\omega} L_x M_x \quad \text{and} \quad R_{0,nature} = \sum_{\alpha}^{\omega} \ell_x m_x$$

Either c_i or v_i' may also be said to represent the ecological selection intensity at age i.

Under biotic restraints, the expressed reproductive span may be shifted later in life relative to the M_x curve, as shown in Figure 1-2. Now negative m_i' and possibly negative c_i arise, and these correct the $\sum_{x=0}^{\omega} c_x$ for the shifted reproductive interval. Figure 1-2 also shows a hatched region depicting the distribution of the missing portion of the $R_{0,max}$ which is estimated by $\sum_{x=0}^{\omega} c_x$. For later conclusions it is necessary to know that negative c_i, representing relaxed selection at age i, can occur.

The definitions of L_x and M_x, and the consequent definition of $R_{0,max}$, used in the above derivation constitute one possible ecological scenario—one which represents the meeting of population and environment at the leve of physiological adaptation. If selection assembles the "best" genotypic combination or combinations, that is, those approaching L_x and M_x in performance, the resulting organisms have maximal adaptation to the nonliving or physical-chemical environment. With biotic interactions negligible this version of the regulative value model seems most applicable to the early evolution of life history pattern in a colonizing population, or life history evolution accompanying a geographical range extension. These definitions of L_x and M_x will be referred to as the "first definitions." In the next section these definitions are changed.

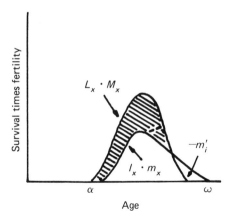

Figure 1-2. Graphical representation of the unrealized portion of net reproduction (hatched area) resulting from less than maximal survival and fertility. The dashed line shows the region of negative m_i' reflected into the hatched region after the manner in which this is accomplished by $\sum_{0}^{\omega} c_x$ when such negative m_i' occur as a result of slowed development and prolongation of life span under ecological or genetic constraints.

Combined Ecological and Genetic Aspects of Selection

In carrying c_i or v_i' over into a model of selection more generality is achieved by redefining L_x and M_x so that they apply to the phenotype or phenotypes with maximal ℓ_x and m_x in the full E_1. This redefinition means that the best performing female phenotypes at any given age contend with the prevailing abiotic and biotic characteristics of the environment E_1 simultaneously. Call this the "second definition" of L_x and M_x. If the biotic interactions of the population are not negligible, L_x and M_x, and consequently $R_{0,max}$, will be lower than under the first definition and maximal physiological adaptation to E_1 may not be possible. Conversely, intense physiological adaptation at some ages may conflict with full adaptation to biotic selective factors.

Either c_i or v_i' can represent the age-specific ecological potential for change generally under natural selection, a potential that in turn shapes the evolution of the life history pattern itself. A high value of c_i or v_i' represents a large ecological potential for change in life history and other associated heritable characters, whereas low, zero, or negative values indicate a restriction, halting, or relaxation of selection. By definition, nonzero c_i are only possible if survival, fertility, or both are variable at age i. If survival, fertility, or both are also heritable at age i, the combining of h_{ik}, the heritability for the kth life history trait at age i, into the quantities $c_i h_{ik}$ or $v_i' h_{ik}$ represents respectively the absolute or relative effectiveness of selection. This measure of age-specific "selection value" is complete if the expression of survival and fertility at age i are independent of the expression of both these traits at all other ages, i.e., there is no trade-off or covariation among life history characters across the ages. Of the two measures of selection value, $c_i h_{ik}$ is perhaps the most useful because it carries the units of net reproduction in c_i. It tells how much larger the net reproductive rate of the population will be when all the heritable variation in fitness character k at age i is exhausted under selection.

When there is covariance between traits across ages, say, when survival at one age and survival or fertility at another age are correlated, an expression for this more complicated form of selection value is:

$$s_{ik} = h_{ik} \sum_m \sum_x c_x r_{ix,m}, \qquad r_{ii,k} = 1$$

for all x (ages) including i, and all life history traits, m, including k. Here the narrow-sense heritability $h_{ik} = V_{Aik}/V_{Pik}$ is defined as usual as the ratio of the appropriate additive genetic and phenotypic variances, and $r_{ix,m}$ is the phenotypic correlation of the mth trait (including trait k) at any age x with the kth trait at age i. Thus, the double summation collects selection values and possible trade-offs between ages and traits as they bear on selection at age i. If linear dependence of traits between ages is not an inadequate description, $r_{ij,m}$ has to be replaced with an appropriate function. The distribution of s_{xk} is meant to describe the entire process of selection created by and acting upon age-specific survival and fertility. Values of $r_{ij,m}$ range from –1 to +1, and some c_x may be negative. Therefore, the summation term could be overall positive or negative and could enhance or reverse the extent of selection indicated by any single value of $c_i h_{ik}$.

Starting with any generation characterized by the regulative value model, ensuing natural selection raises the mean value of survival and fertility at each age i in proportion to the s_{ik}. Fertility, of course, can rise only if age i is a reproductive age. Concurrently, expressed R_0 will rise and the genetic variability for survival and fertility will decline. In a simple reproductive sense, phenotypes with values above the current average for trait k at age i will recur more frequently than phenotypes below the mean. This unequal representation of values will change the age-specific values of ℓ_x and m_x so that:

$$\ell_i \rightarrow (\ell_i + \ell'_{ii}) \qquad \text{and} \qquad m_i \rightarrow (m_i + m'_i)$$

For this process to run to completion sufficient time would be required for recombination to elevate the frequencies of the more favored and possibly rarer genotypes. The decay of polygenic variation for survival and fertility and the consequent decline of the c_i and $\sum_{x=0}^{\omega} c_x$ under directional selection warrant careful study because they control the rate of change in life history pattern.

If the s_{ik} were negative frequently enough, expressed R_0 would decrease and extinction eventually ensue. This would represent too great a mismatch between the available genetic structure of the population and the surrounding ecological conditions, a sort of dysgenic life history arising from extreme negative, but heritable, phenotypic correlations.

Although most would agree that patterns of survival and fertility are essential aspects of ecologically realistic models of life history evolution, and of selection models generally, it is fair to ask whether the foregoing model can be extended to represent the evolution of characters other than age-specific survival and fertility. The answer is yes, if the genetic covariation of another trait with survival or fertility, or both, is known and if the phenotypic variance of the other trait is known. Then it should be possible to follow the additional trait, be it a life history or a behavioral or morphological character, as a correlated response (Falconer 1960) to the selective modification of ℓ_x and m_x.

Mesoevolutionary Models of Life History Change

The demographic model of the last section develops a possible kernel for more complex models where longer term transformations of life history pattern are driven not only by selection in the microevolutionary sense, but also by major environmental shifts which periodically reset the conditions for microevolution. For example, a sequence of environment episodes . . . $\rightarrow E_1 \rightarrow E_2 \rightarrow E_3 \rightarrow$ etc., can be imagined with the successive environments varying both in temporal duration and the way they initially set the vital statistics of a population and their heritable expression. If it is supposed that the environmental sequence runs for hundreds or thousands of generations, the problem of realism on the scale of mesoevolutionary models of life history evolution is encountered. It is clear that population size and the combined effect of the ecological setting and additive genetic variation for survival and fertility remain essential properties of the model. Using a model such as the regulative value approach will tell something of changes within any one environmental episode and whether extinction is likely or not to ensue in that episode. Missing, however, are relationships that map the microevolutionary process recursively through major environmental shifts and major alterations of life history pattern and attendant genetic structure. If large and cumulative changes of this sort could be represented, the beginnings of a quantitative mesoevolutionary theory would be obtained. Life history evolution seems deeply implicated, but the pieces of the puzzle are not obvious. It seems to me that only with such models will there be any hope of an adequate theoretical exploration of such big evolutionary phenomena as adaptation and sexuality, speciation, species selection, extinction, and large-scale morphological divergence.

The relation between generation time and the duration of any environmental episode is a particularly intriguing one. If the episode is too short, two few generations will pass to allow the life history pattern and its genetic structure to settle to equilibrium. Hence it is possible, that an evolving population may be so frequently subject to environmental shifts that its life history pattern seldom reflects any sort of "optimal" or even available (Stearns 1976) solution—a sort of mesoevolutionary "r selection." Conversely, populations waiting long in some consistent environment should stabilize in life history and genetic structure, except for the presumably slow change supported by mutation. This is not to say that such a stabilized condition necessarily leaves the population without some heritable variation for life history characters, at least in fluctuating environments (Istock 1978, 1981a,b, Lande 1976).

Conclusion

The principal suggestions about life history evolution made here are that such evolutionary changes typically depend on a polygenic mode of inheritance, derive first of all from age-specifically arrayed birth and death events, and that the evolution of statistically significant changes in life history pattern may often be extensive enough, in a genetic sense, to constitute mesoevolution: the scale of evolutionary change where irreversibility is achieved.

Realism in models of life history evolution demands a degree of ecological complexity not faced in earlier models of evolution. In particular, definitions of environment, the characterization of genetic structure, and the extensiveness of the suite of life history characters incorporated in a model pose formidable problems of dimensionality. Processes of physiological and geographical adaptation and species interaction and coevolution add to the potential complexity of such models.

Among the details of the model of life history evolution presented here, the incorporation of a measure of the age-specific retardation of population growth, c_i, brought out two different definitions of the phenotypic potential for evolution. The first definition ignores the effect of biotic interactions on the microevolution of a life history pattern and may be appropriate in some simple cases of physiological adaptation to a new climate or the extension of the geographical range of a species. However, even in cases of physiological adaptation and range extension the second definition of a phenotypic potential for life history evolution, leading to a different set of c_i, may be the most appropriate theoretical object. Under the second definition, the potential for life history evolution incorporates biotic and abiotic influences together.

The distinction of abiotic and biotic factors is a time-honored one in ecology, but it takes on further meaning in studies of life history evolution. The biotic components of the selective environment are themselves other species populations and liable to evolve. Hence, even though some environment may be consistent in its physical and chemical features over a period of time, the biotic components may constantly or frequently change. I have not attempted to meet this difficulty; the regulative value model assumes that the pattern of biotic effects distributed over a life history does not change except as the population adapts to reduce their depressing influence on the net reproductive rate.

Based on many ecological studies it is reasonable to expect some biotic components to be density-dependent. The coupled evolution of two or more species through reciprocal density-dependent relations, as, say, in the case of competing or mutualistic species, adds a further aspect to evolutionary irreversibility. Now the ecological setting in which genetic change may reverse requires a pretty carefully contrived, and probably unlikely, reversal of multiple species interrelations. It is known that coevolutionary phenomena are quite common in nature; hence, the complexity of ecological circumstance eventually assists in the enforcement of genetic irreversibility.

Life history evolution, as represented here, is consonant with both sexuality and asexuality and highlights the importance of understanding the relative potentials for life history evolution in both sexual and obligately asexual forms. Each generation, selection scans the expressions of life history within an entire population of some sexual form, and varying expressions are weighted in units of reproduction for their intrinsic ecological and genetic "merit." At a given age higher fertility is found, at another age a higher probability of survival has found expression, and the two high performances may or may not be phenotypically and genetically correlated. Sexuality and genetic recombination, through additive polygenic variation, will fashion many new combinations from these partial successes. For asexual forms, the scanning by natural selection creates a winnowing of clones that vary by dint of mutation or separate origins from sexual ancestors.

Acknowledgments. I am deeply grateful to Tom Caraco, William Etges, Kathryn Murphy, John Moeur, Uzi Nur, Joseph Warnick, and Loy Merkle for their many comments on the ideas in this chapter and for critical comments on an earlier draft of the manuscript. I am indebted to Millie Grassi and Sandra Wright for the typing of the manuscript. This work was supported by grants from the National Science Foundation.

References

Bell, G.: On breeding more than once. Am. Nat. 110, 57-77 (1976).

Cole, L. C.: The population consequences of life history phenomena. Q. Rev. Biol. 29, 103-137 (1954).

Dobzhansky, T.: Genetics and the Evolutionary Process. New York: Columbia University Press, 1970.

Falconer, D. S.: Introduction to Quantitative Genetics. New York: Ronald Press, 1960.

Fisher, R. A.: The Genetical Theory of Natural Selection. New York: Dover, 1958.

Gould, S. J.: Is a new and general theory of evolution emerging? Paleobiology 6, 119-130 (1980).

Istock, C. A.: Natural selection in ecologically and genetically defined populations. Behav. Sci. 15, 101-115 (1970).

Istock, C. A.: Fitness variation in a natural population. In: Evolution of Insect Migration and Diapause. Dingle, H. (ed.). New York: Springer-Verlag, 1978.

Istock, C. A.: The extent and consequences of heritable variation for fitness characters. In: Population Biology: Retrospect and Prospect. Oregon State Colloquium in Biology. King, C. R., Dawson, P. S. (eds.). New York: Columbia University Press, 1981 a, in press.

Istock, C. A.: Natural selection and life history variation: Theory plus lessons from a mosquito. In: Insect History Patterns. Denno, R., Dingle, H. (eds.), pp. 113-127. New York: Springer-Verlag, 1981 b.

Istock, C. A., Tanner, K., Zimmer, H.: Habitat selection in the pitcher-plant mosquito, *Wyeomyia smithii*: Behavioral and genetic aspects. In: Phytolemata: Terrestrial Plants as Hosts of Aquatic Insect Communities. Lounibos, P., Frank, H. (eds.). Marlton: Tlexus, 1982, in press.

Lande, R.: The maintenance of genetic variability by mutation in a polygenic character with linked loci. Genet. Res. Cambr. 26, 221-235 (1976).

Levene, H.: Genetic equilibrium when more than one niche is available. Am. Nat. 87, 331-333 (1953).

Lewontin, R. C.: Fitness, survival and optimality. In: Analysis of Ecological Systems. Horn, D. J., Stairs, G. R., Mitchell, R. D. (eds.), pp. 3-21. Columbus: Ohio State University Press, 1979.

Lewontin, R. C.: Fitness, survival and optimality. In: Analysis of Ecological Systems. Horn, D. J., Stairs, G. R., Mitchell, R. D. (eds.). Columbus: Ohio State University Press, 1979.

Murphy, G. I.: Patterns in life history and environment. Am. Nat. 102, 391-403 (1968).

Saul, S. H.: Fitness, adaptation and interdemic selection in populations of *Drosophila*. Doctoral Dissertation, University of Rochester, New York. Ann Arbor: University Microfilms, (71-01427) 1970.

Schaffer, W. M.: Optimal reproductive effort in fluctuating environments. Am. Nat. 180, 783-790 (1974).

Stanley, S. M.: Macroevolution: Pattern and Process. San Francisco: W. H. Freeman, 1979.

Stearns, S. C.: Life history tactics: A review of the ideas. Q. Rev. Biol. 51, 3-47 (1976).

Stearns, S. C.: The evolution of life history traits. Ann. Rev. Col. System. 8, 145-172 (1977).

Tauber, M. J., Tauber, C. A.: Evolution of phenological strategies in insects: A comparative approach with eco-physiological and genetic considerations. In: Evolution of Insect Migration and Diapause. Dingle, H. (ed.), pp. 53-71. New York: Springer-Verlag, 1978.

Thoday, J. M.: Natural selection and biological progress. In: A Century of Darwin. Barnett, S. A. (ed.), pp. 313-323. London: Heineman, 1958.

Tuljapurkar, S. D., Orzack, S. H.: Population dynamics in variable environments. I. Long-run growth rates and extinction. Theor. Pop. Biol. 18, 314-342 (1980).

Chapter 2

Elements of a Quantitative Genetic Model of Life History Evolution

Russell Lande*

Introduction

Much of the theory of life history evolution is based on optimization models that attempt to predict the equilibrium state(s) of a population by maximizing its growth rate, subject to ad hoc constraints and trade-offs between individual growth, reproduction, and survival (Stearns 1977). However, the types of evolutionary processes that produce life histories that are optimal in some sense can only be determined from genetic models of selection in age-structured populations (Charlesworth 1980). For example, frequency-dependent selection can produce maladaptive evolution, decreasing the mean fitness in a population (Wright 1969, Chapter 5). Despite their greater complexity, genetic models incorporating evolutionary constraints in measurable patterns of genetic variation and natural selection have the advantage of providing a dynamic rather than a static description of evolution and provide a framework for quantitative testing of hypotheses about constraints on life history evolution.

Two fundamental requirements for a robust science of life history evolution are (1) experimental methods of measuring heritable variation in life history characters within populations, and (2) a dynamical theory describing how this variation evolves. Estimates of genetic variance and covariance in quantitative characters can be obtained from phenotypic correlations between relatives or from artificial selection experiments (Falconer 1960, Kempthorne 1957, Mode and Robinson 1959). The same methods can be extended to life history characters measured at specific ages. Using these techniques, Rose and Charlesworth (1980, 1981a, b) estimated strong negative genetic correlations of early fecundity in females with late fecundity and life span but were unable to detect any increase with age in additive or dominance genetic variance for age-specific fecundity. Their findings support the hypothesis that senescence is caused by genes with beneficial effects early in life and deleterious effects in old age (Williams 1957) and do not support an alternative hypothesis that senescence is largely a result

*Department of Biophysics and Theoretical Biology, University of Chicago, 920 East 58th Street, Chicago, Illinois 60637 U.S.A.

of mutations that have only detrimental effects, expressed late in life (Hamilton 1966). Data of Sacher and Duffy (1979) indicate that longevity in male mice is determined primarily by metabolic and morphological characters with an intermediate optimum, rather than deleterious genes expressed only in old age, suggesting that most of the genetic variance in longevity is nonadditive (see Wright 1935).

Although experimental techniques for measuring genetic constraints on life history evolution already exist, at present there is no dynamical theory of natural selection on quantitative characters in age-structured populations. Some major questions that such a theory should address are: (1) What is the evolutionary significance of genetic variance and covariance in life history traits? (2) Is there a general principle of adaptation for correlated characters in age-structured populations? (3) Can information on genetic variance and covariance of life history characters be used to draw inferences about the forces of natural selection affecting them?

This chapter outlines some elements of a dynamic theory of life history evolution by combining standard models of quantitative genetics and demography for populations with discrete, overlapping generations. Further discussion of these and other topics, using demographic models with continuous time, can be found in Lande (1982).

Elements of a Dynamic Theory

For simplicity in describing the evolutionary influence of overlapping generations, the following models concern populations that are monoecious, or that have identical selection pressure on both sexes and no sexual dimorphism. Consider a vector of quantitative characters of an individual,

$$z = \begin{pmatrix} z_1 \\ \vdots \\ z_n \end{pmatrix} \tag{2-1}$$

some or all of which may be age specific. As frequently can be arranged by a simple scale transformation, most often logarithmic, the joint distribution of the characters in a population of unselected individuals is assumed to be multivariate normal (Gaussian), with mean $\bar{z}$ and phenotypic variance-covariance matrix $\mathbf{P}$. Observations on phenotypic correlations between relatives, or artificial selection experiments, can be employed to partition $\mathbf{P}$ into additive genetic and environmental (plus nonadditive genetic) components. If there are no genotype-environment correlations the additive genetic variance-covariance matrix, $\mathbf{G}$, and the environmental variance-covariance matrix, $\mathbf{E}$, sum to

$$\mathbf{P} = \mathbf{G} + \mathbf{E} \tag{2-2}$$

It is also assumed that the distribution of additive genetic effects (breeding values) for the characters is multivariate normal, as will tend to apply for polygenic characters with a Gaussian phenotype distribution.

The basic demographic model of age-structured populations is utilized (Goodman 1968, Leslie 1945, 1948). However, the age-specific mortality and fecundity rates are allowed to depend on the quantitative characters that compose the phenotype of an individual during its lifetime. One assumption necessary to render the analysis tractable is that selection is weak, i.e., that all life history phenotypes in a population have similar vital statistics. Then life history evolution is slow in comparison to normal demographic processes and, except for a few generations just after its establishment, a population evolves within a slowly changing "stable" age distribution determined by the average age-specific mortality and fecundity rates.

The population is divided into discrete age classes, evenly spaced in time $0, 1, \ldots, L$, where L is the maximum life span attainable in the population. The probability of survival of an individual to age a is l_a and its expected fecundity at that age is m_a. Under some mild conditions on the average l_a and m_a schedules (see Leslie 1945), the proportion of individuals of age a in the population at the "stable" age distribution is

$$\lambda^{-a} \bar{l}_a \left/ \sum_{a=0}^{L} \lambda^{-a} \bar{l}_a \right. \tag{2-3}$$

where λ is the geometric rate of growth of the population per unit time, obtained as the unique positive real root of the equation

$$\sum_{a=0}^{L} \bar{w}_a \lambda^{-a-1} = 1 \tag{2-4}$$

with

$$\bar{w}_a = \overline{l_a m_a} .$$

$\bar{w}_a$ may be conceived as the mean age-specific fitness, the population average of viability to age a times fecundity at that age. The proportion of individuals of age 0 contributed by parents of age a is, using Eq. (2-4),

$$q_a = \lambda^{-a} \bar{w}_a \left/ \sum_{a=0}^{L} \lambda^{-a} \bar{w}_a \right.$$

$$= \lambda^{-a-1} \bar{w}_a \tag{2-5}$$

Consider first the case where the vector of characters of an individual, **z**, does not change with age. The basic theory of selection on a quantitative character in an age-structured population (Dickerson and Hazel 1944, Rendel and Robertson 1950) asserts that under a constant regime of age-specific selection with a stable age distribution and genetic variation constant in time, the rate of response in the mean phenotype per unit time in any age class is equal to the average genetic selection differential for parents of all ages, divided by the average age at reproduction (see also Hill 1974). The generalization to a vector of characters is straightforward. Under weak selection and slow phenotypic evolution, the age-specific selective forces are nearly constant during a generation. The rate of evolution of the mean phenotype per unit time is approximately

$$\Delta \bar{z} = T^{-1} \sum_{a=0}^{L} q_a \Delta \bar{g}_a \tag{2-6}$$

where

$$T = \sum_{a=0}^{L} (a + 1)q_a \tag{2-7}$$

is the generation time, the average age at reproduction, and $\Delta \bar{g}_a$ is the vector of genetic selection differentials at age a, the difference in the mean breeding values of selected and unselected individuals at that age.

Because the rate of response in the vector of mean breeding values under weak selection is approximately the same for all age classes, these formulas also describe the evolution of the mean life history in the entire population. It is apparent, therefore, that some or all of the characters in the vector z may be regarded as age specific (expressed at certain ages) with the implied stipulation that viability and fecundity selection on individuals of age a can operate directly only on characters expressed at that age. Genetic correlations between characters expressed at different ages will nevertheless produce correlated selective responses between them.

The vector of genetic selection differentials at age a caused by variation in individual viabilities and fecundities, assuming these are not frequency dependent, can be written from formulas of Lande (1979) as

$$\Delta \bar{g}_a = \bar{w}_a^{-1} \mathbf{G} \nabla \bar{w}_a \tag{2-8}$$

where $\mathbf{G}$ is the additive genetic variance-covariance matrix of the characters in unselected individuals and

$$\nabla = \begin{pmatrix} \partial/\partial \bar{z}_1 \\ \vdots \\ \partial/\partial \bar{z}_n \end{pmatrix}$$

is the gradient operator with respect to changes in the mean phenotypes. Substituting Eq. (2-8) into Eq. (2-6), employing Eq. (2-5), produces

$$\Delta\bar{z} = T^{-1} \mathbf{G} \sum_{a=0}^{L} \lambda^{-a-1}\nabla\bar{w}_a \qquad (2\text{-}9)$$

Applying the gradient operator to Eq. (2-4), using Eqs. (2-5) and (2-7), gives

$$\sum_{a=0}^{L} \lambda^{-a-1}\nabla\bar{w}_a = \sum_{a=0}^{L} \bar{w}_a(a+1)\lambda^{-a-2}\nabla\lambda$$

$$= T\lambda^{-1}\nabla\lambda = T\nabla\ln\lambda \qquad (2\text{-}10)$$

where $\ln\lambda$ is the natural logarithm (base e) of λ. Equations (2-9) and (2-10) reveal a simple dynamical law,

$$\Delta\bar{z} = \mathbf{G}\nabla\ln\lambda \qquad (2\text{-}11)$$

If the mortality and fecundity schedules of individual life history phenotypes, and the pattern of genetic variation among them, remain constant in time it can be shown from Eq. (2-11) using methods in Lande (1979) that under weak selection the evolution of the mean life history continually increases the growth rate of the population, until an equilibrium is reached,

$$\Delta\lambda \geqslant 0 \qquad (2\text{-}12)$$

Under conditions of density-independent population growth, an equilibrium of the mean life history occurs at a local maximum of the population growth rate, provided the additive genetic variance-covariance matrix is not singular, $|\mathbf{G}| \neq 0$. With density-dependent population growth λ must usually be near unity, and Fisher (1958, Chapter 2) and Charlesworth (1980, Chapter 5) show that in some cases life history evolution maximizes the equilibrium population size, $\hat{N}$, rather than λ.

Therefore, λ (or $\hat{N}$), as a function of the mean life history phenotypes, $\bar{z}_1, \ldots, \bar{z}_n$, can be viewed as an adaptive topography for phenotypic evolution in an age-structured population. The concept of an adaptive topography for phenotypes is discussed extensively in qualitative terms by Simpson (1953), by analogy with Wright's (1932, 1969) adaptive surface for gene frequencies and Fisher's (1958) fundamental theorem of natural selection. Genetic evolution in a given environment increases the level of adaptation of a population, whereas changes in the physical and biotic factors that alter the adaptive topography tend to decrease adaptation (Fisher's "deterioration" of the environment).

It is important to note that whereas the joint evolution of all life history characters in a constant environment increases the growth rate of the population, no such principle applies to any subset of characters that is genetically correlated with other selected traits. The *selection gradient*, $\nabla\ln\lambda$, represents the forces of directional selection on the characters, because the element $\partial\ln\lambda/\partial\bar{z}_i$ is the change in the logarithmic growth rate of the population caused by a small change in $\bar{z}_i$, holding all other $\bar{z}_j$ constant. The rate and direction of life history evolution are not determined by the selective forces alone but depend critically on the additive genetic variance-covariance structure of the characters. The net rate of response in a particular trait to the overall pattern of selection is a sum of the direct response to selection on the trait, and the correlated responses to selection on genetically correlated characters (Eq. 2-11),

$$\Delta \bar{z}_i = \sum_{j=1}^{n} G_{ij} \, \partial \ln \lambda / \partial \bar{z}_j$$

where G_{ij} is the ijth element of the matrix $\mathbf{G}$.

At any given time the direction of evolution of a particular character may be against the actual force of selection on it, because of genetically correlated responses to selection on other characters. The adaptive response to selection on the whole organism and its life history therefore may involve compromises between characters that are genetically correlated. A compromise of this kind occurs when natural selection acts to increase each of a set of characters with negative genetic correlations. Then any progress in one trait results in a correlated decrease in the others, known as "genetic slippage" (Dickerson 1955). With sufficiently large negative correlations between characters, it may be impossible for selection simultaneously to increase all of them. This situation may often arise for major components of fitness, such as fecundity and viability, even though the phenotypic characters that determine these fitness components retain substantial additive genetic variance (Dickerson 1955, Falconer 1960, Chapter 20). Sustained directional selection on major components of fitness tends to rapidly fix pleiotropic (or linked) mutations with positive effects on the components and to leave segregating those mutations with opposite effects on the components of fitness. For example, in a flock of chickens at a limit to artificial selection for several traits, including egg size and egg number, Dickerson (1955) measured a large negative genetic correlation and a phenotypic correlation near zero for these two traits, indicating a positive environmental correlation. Within populations of rye grass at a limit to artificial selection for increased yield, Cooper and Edwards (1961) estimated large negative genetic correlations and positive environmental correlations between tiller number and dry weight per tiller, and between leaf size and rate of leaf production. The pattern of phenotypic correlations between major components of fitness often does not resemble the underlying pattern of genetic constraints on the evolution of the mean life history in a population.

Characters other than major components of fitness are most likely to be under natural selection toward an intermediate optimum phenotype. For such characters, the assumption that the phenotypic and additive genetic variation parameters, $\mathbf{P}$ and $\mathbf{G}$, remain constant during the evolution of the mean phenotypes is often a good approximation for several generations even after the onset of intense artificial selection (Falconer 1960). In natural populations, where genetic variation is maintained by a continual flux of spontaneous polygenic mutation and the pattern of stabilizing selection around a (moving) optimum phenotype may be very conservative, $\mathbf{P}$ and $\mathbf{G}$ may remain nearly constant during long periods of time (Lande 1975, 1980). The temporal and taxonomic limits within which this assumption is valid can be ascertained empirically by comparing phenotypic and genetic variation patterns in related populations. For instance, Lande (1979) inferred substantial constancy of variation parameters of adult brain and body weights among insectivores and rodents, but with rather different values in primates, and Arnold (1981) measured similar patterns of genetic variation in feeding preferences in divergent races of garter snakes.

Equation (2-11) can be combined with information on additive genetic variance and covariance among characters to analyze the forces of natural selection in age-structured populations. If **G** remains constant in time, although the selective forces may be slowly fluctuating, the net selection gradient required to produce an observed change in the mean life history of a population can be calculated by summing Eq. (2-11) through the appropriate span of time and inverting **G**, after eliminating linear combinations of characters for which there is no additive genetic variance so that **G** is not singular.

$$\sum_t \nabla \ln \lambda = \mathbf{G}^{-1} \left[\bar{z}(t) - \bar{z}(0) \right]$$

This method also can be extended to deduce the direction of past selective forces that have acted to differentiate contemporary populations descended from a common ancestral population (Lande 1979). The use of this formula requires evidence that the phenotypic changes being analyzed result from genetic evolution and are not caused by immediate effects of the environment. It is sometimes feasible to account for immediate environmental effects by raising samples from the populations in a common environment.

For species in which individuals grow to a final adult size and shape, it is especially important to know the conditions under which the evolution of the adult phenotype can be analyzed independently of preadult characters, when the former is of primary interest, because this reduces the number of characters involved. This simplification is valid if (viability) selection on embryonic and juvenile characters is negligible, or if these are genetically uncorrelated with adult characters. The evolution of the adult phenotype can then be analyzed as an autonomous system.

It is often convenient to analyze life history variation and evolution by considering homologous traits at different ages as separate characters with genetic and phenotypic correlations, e.g., age-specific body size or growth rates. Organisms with complex life cycles, such as many arthropods and amphibians, naturally lend themselves to such an approach. However, if several characters at many ages are considered, the analysis may become intractable in practice because of the large number of variation parameters that must be estimated. For organisms with continuous growth and shape change in development, a powerful method of reducing the number of characters is the analysis of parameters of individual growth curves. In many species of plants, reptiles, birds, and mammals, the observable ontogeny of an individual can be described by a few constant parameters, such as allometric coefficients, growth rates, and initial and final size and shape. Life history variation and evolution then can be analyzed in terms of these growth curve parameters (Atchley and Rutledge 1980, Cock 1966, Kidwell et al. 1952, 1979).

Acknowledgments. I thank J. Antonovics, S. J. Arnold, B. Charlesworth, A. R. Kiester, and E. G. Leigh, Jr., for helpful discussions and criticisms. This work was partially supported by U. S. Public Health Service Grant GM27120-01.

References

Arnold, S. J.: Behavioral variation in natural populations. I. Phenotypic, genetic and environmental correlations between chemoreceptive responses to prey in the garter snake, *Thamnophis elegans. Evolution* 35, 489-509 (1981).

Atchley, W. R., Rutledge, J. J.: Genetic components of size and shape. I. Dynamic components of phenotypic variability and covariability during ontogeny in the laboratory rat. *Evolution* 34, 1161-1173 (1980).

Charlesworth, B.: *Evolution in Age-Structured Populations.* Cambridge: Cambridge University Press, 1980.

Cock, A. G.: Genetical aspects of metrical growth and form in animals. *Q. Rev. Biol.* 41, 131-190 (1966).

Cooper, J. P., Edwards, K. J. R.: The genetic control of leaf development in *Lolium. Heredity* 16, 63-82 (1961).

Dickerson, G. E.: Genetic slippage in response to selection for multiple objectives. *Cold Spring Harbor Symp. Quant. Biol.* 20, 213-224 (1955).

Dickerson, G. E., Hazel, L. N.: Effectiveness of selection on progeny performance as a supplement to earlier culling in livestock. *J. Agr. Res.* 69, 459-476 (1944).

Falconer, D. S.: *Introduction to Quantitative Genetics.* New York: Ronald Press, 1960.

Fisher, R. A.: *The Genetical Theory of Natural Selection.* (2nd ed.) New York: Dover, 1958.

Goodman, L. A.: An elementary approach to the population projection matrix, to the population reproductive value, and to related topics in the mathematical theory of population growth. *Demography* 5, 382-409 (1968).

Hamilton, W. D.: The moulding of senescence by natural selection. *J. Theor. Biol.* 12, 12-45 (1966).

Hill, W. G.: Prediction and evaluation of response to selection with overlapping generations. *Anim. Prod.* 18, 117-139 (1974).

Kempthorne, O.: *An Introduction to Genetic Statistics.* Ames: Iowa State University Press, 1969.

Kidwell, J. F., Gregory, P. W., Guilbert, H. R.: A genetic investigation of allometric growth in Hereford cattle. *Genetics* 37, 158-174 (1952).

Kidwell, J. F., Herbert, J. G., Chase, H. B.: The inheritance of growth and form in the mouse. V. Allometric growth. *Growth* 43, 47-57 (1979).

Lande, R.: The maintenance of genetic variability by mutation in a polygenic character with linked loci. *Genet. Res.* 26, 221-235 (1975).

Lande, R.: Quantitative genetic analysis of multivariate evolution, applied to brain: body size allometry. *Evolution* 33, 402-416 (1979).

Lande, R.: The genetic covariance between characters maintained by pleiotropic mutations. *Genetics* 94, 203-215 (1980).

Lande, R.: A quantitative genetic theory of life history evolution. *Ecology* (in press).

Leslie, P. H.: On the use of matrices in certain population mathematics. *Biometrika* 33, 183-212 (1945).

Leslie, P. H.: Some further notes on the use of matrices in population mathematics. *Biometrika* 35, 213-245 (1948).

Mode, C. J., Robinson, H. F.: Pleiotropism and the genetic variance and covariance. *Biometrics* 15, 518-537 (1959).

Rendel, J. M., Robertson, A.: Estimation of genetic gain in milk yield by selection in a closed herd of dairy cattle. *J. Genet.* 50, 1-8 (1950).

Rose, M., Charlesworth, B.: A test of evolutionary theories of senescence. *Nature* (London) 287, 141-142 (1980).

Rose, M. R., Charlesworth, B.: Genetics of life-history in *Drosophila melanogaster*. I. Sib-analysis of adult females. *Genetics* 97, 173-186 (1981 a).

Rose, M. R., Charlesworth, B.: Genetics of life-history in *Drosophila melanogaster*. II. Exploratory selection experiments. *Genetics* 97, 187-196 (1981 b).

Sacher, G. A., Duffy, P. H.: Genetic relation of life-span to metabolic rate for inbred mouse strains and their hybrids. *Fed. Proc.* 38, 184-188 (1979).

Simpson, G. G.: *The Major Features of Evolution*. New York: Columbia University Press, 1953.

Stearns, S. C.: The evolution of life history traits: A critique of the theory and a review of the data. *Ann. Rev. Ecol. Syst.* 8, 145-171 (1977).

Williams, G. C.: Pleiotropy, natural selection, and the evolution of senescence. *Evolution* 11, 398-411 (1957).

Wright, S.: The roles of mutation, inbreeding, crossbreeding and selection in evolution. *Proc. Sixth Intl. Congress Genet.* 1, 356-366 (1932).

Wright, S.: The analysis of variance and the correlations between relatives with respect to deviations from an optimum. *J. Genet.* 30, 243-256 (1935).

Wright, S.: *Evolution and the Genetics of Populations. Vol. 2. The Theory of Gene Frequencies*. Chicago: University of Chicago Press, 1969.

PART TWO

Physiological Adaptation

One way that selection can act on life histories and contribute to Istock's regulative value is through adaptation to abiotic factors. Two factors in the physical environment that are of major importance to a wide array of organisms are temperature and photoperiod. Brian Bradley considers the first with respect to copepods in a seasonal estuarine environment, and Catherine and Maurice Tauber consider interactions between both in the evolution of phenological patterns in a group of predatory insects, the green lacewings (*Chrysopa*). These patterns include both diapause and differences in developmental rates.

Bradley examines physiological and genetic variance in temperature tolerance in *Eurytemora*, a copepod naturally exposed to wide temperature fluctuations. Substantial genetic variance for tolerance exists in spite of pronounced individual physiological flexibility. Current models seem to Bradley inadequate to account for this unless the genetic variance reflects polygenic mutations. A combined model merging notions of disruptive selection and genetic homeostasis leads to the conclusion that the maintenance of genetic correlations among life history traits and physiological flexibilities may be of paramount importance. Because genetic correlations are additive genetic covariances standardized relative to additive genetic standard deviations, Bradley lends empirical reinforcement to Lande's emphasis on additive genetic variance-covariance relationships among life history traits.

An effective approach to the relation between seasonality and life history adaptations is comparison among geographical populations. The Taubers examine samples of a species complex of *Chrysopa* from an array of locales across North America. These populations display local adaptation of coordinated sets of life history traits in response to prevailing conditions of temperature and photoperiod. Much of the observed variation is continuous and apparently polygenic, but some, especially important patterns of voltinism which reproductively isolate closely related sympatric species, seem to be a function of a relatively simple Mendelian system acting in crosses between species. The expression of genetic variation influencing *Chrysopa* life histories ranges from continuous variation in metric characters to polymorphism and finally to the full life cycle asynchrony in potentially interbreeding species, suggesting that irreversible change (mesoevolution) has occurred.

Chapter 3

Models for Physiological and Genetic Adaptation to Variable Environments

BRIAN P. BRADLEY*

Introduction

The challenge of a temporally or spatially varying environment to a population can be met at the individual level, at the population level, or at both. Individual organisms may adjust physiologically to the entire range of conditions. Populations may systematically change in gene frequencies (where the environmental variation is sufficiently predictable). Likewise, both physiological and genetic changes may occur in space and time.

In the case of the copepod *Eurytemora affinis* found throughout the year in the Chesapeake Bay region, populations are exposed to temperatures from near 0°C to 30°C and above, and individuals up to a 10°C range in temperature. Generation time varies from 3 months at the lower temperature to 10 days at the highest temperature. Temperature tolerance is therefore an important trait in *Eurytemora*, and both physiological and genetic variance in temperature tolerance have been observed repeatedly (Bradley 1978a, b).

If, because of this physiological variance, individual copepods, that is to say particular genotypes, could survive and reproduce in the entire range of temperatures, the question would be why so much genetic variance remains (Slobodkin and Rapoport 1974). If the genetic variance is selectively maintained, and the neutralist hypothesis is surely irrelevant in the case of temperature tolerance, then the inquiry becomes which form of selection is most likely.

Selection models including an advantage to heterozygotes and temporal and spatial models of the kind suggested by Haldane and Jayakar (1963) and by Levene (1953), respectively, are obvious possibilities. Each of these appears to be unlikely in *Eurytemora*, as do models by Ewing (1979) combining spatial and temporal components, in view of the lack of evidence of dominance or of genotype-environment interaction in temperature tolerance. For the same reasons spatial and temporal models by Gillespie (1974a, b, 1976) appear inapplicable to *Eurytemora*. At present there remain two models, both dating from the 1950s, both including simultaneous selection pressure on two traits, which may explain the physiological and genetic variance observed in tem-

*Department of Biological Sciences, University of Maryland, Baltimore, Maryland 21228 U.S.A.

perature tolerance in *Eurytemora*. A third, quite different model proposed by Lande (1976) is also a possibility.

After a description of the experimental and analytical methods used in the study and a presentation of pertinent results, the various models are discussed in turn. Finally a tentative, and I believe testable, model is proposed combining the first two plausible models mentioned above.

The Organism

Eurytemora affinis (Poppe) is a calanoid copepod found predominantly in estuaries in temperate regions but also reported in the Great Lakes (Engel 1962). *Eurytemora* is ideally suited for both laboratory and field studies in population biology. The sexes are distinguishable; the female requires one mating for each viable egg sac produced; generation time is approximately three weeks at 15°C; and cultures are easily maintained in filtered estuary water over many generations in the laboratory.

Methods

Collection

Samples from the wild are collected using a 73-μm mesh plankton net towed behind a small boat. Ovigerous females are isolated from the samples to set up pure cultures.

Maintenance

Cultures of *Eurytemora* can be maintained through many generations in 2- or 4-liter flasks of filtered estuary water from 0 to 7 ppt salinity when fed a mixture of algal species, including at least one diatom. Culture methods were consistent throughout the study and were completely uniform within a particular experiment.

Individual broods are raised by isolating ovigerous females in 125- or 250-ml flasks, returning the female to stock culture, and feeding the released nauplii.

Measurements

The measurements routinely made on *Eurytemora* are shown in Table 3-1. When data are collected on individual broods, population variances in these characters can be partitioned between and within families (single clutches). The emphasis to date has been on temperature tolerance, in particular tolerance to high temperature. Taking temperature tolerance as a measure of survival, the characters in Table 3-1 together represent overall fitness in a copepod population, assuming one egg sac per female. For a complete measure of production, information would also be required on number of egg sacs per female and age at last reproduction, as well as on the mortality of adults. During the stressful (summer) period of the year, the longevity of adults is such that the interval between egg sacs is not a critical variable.

Table 3-1. Traits measured in *Eurytemora affinis*

Temperature tolerance (Preacclimation)	A
Temperature tolerance after acclimation	B
at 23°C for 24 hr[a] (Post acclimation)	
Flexibility (degree of acclimation)	B - A
Egg production per egg sac per female	
Development time (egg to egg = age at	
first reproduction)	
Viability (egg to adult)	
Sex Ratio (No. of females per total adults)	

[a] Or at 4°C for 24 hr in the case of cold tolerance.

The measurement of temperature tolerance currently used was prompted by two reports, one by Battaglia (1967), reporting different rates of recovery from osmotic shock of copepod populations adapted to different salinities, and the other by Gonzalez (1974), showing that critical thermal maximum (CTM), where equilibrium of the animal is lost, and upper lethal temperature (ULT) are related in copepods, at least between locations. The assay that I first used (Bradley 1975) depended on time to succumb (or enter a coma) following temperature shock, together with time to recover. I found that animals became comatose around 35°C no matter what the conditions, so CTM was not a useful measure in *Eurytemora*.

The present assay begins with a temperature shock of 32°C, administered to individual copepods in 10 ml of water in shell vials, by lowering the vials in a rack into an aquarium preheated to 32°C. The temperature is then raised 0.5°C at 5-min intervals, thereby accommodating a wide range of tolerances in a short-term (30-40 min) assay. Temperature increase using a heating-stirring unit is monitored and controlled with a thermistor attached to a PDP-8 laboratory computer. The assay is the elapsed time (minutes) until an individual becomes immobile, even after slight agitation of the vial. This assay has been shown to predict survival quite well, with a correlation between tolerance assayed and survival time at 33°C of 0.83 (p < 0.01) (Bradley 1976). Little information is lost by omitting recovery time because times to succumb and recovery have a high negative correlation.

The assay for cold temperature tolerance consists of immersing the vials in an aquarium at 1.5°C for 10 min, at which time they are removed. Both time to succumb (TS) and time to recover (TR) are noted and the cold tolerance index given as 30 - TS + TR, because the total time of the assay is 30 min. Those animals not becoming comatose before 10 min are given a score of 40.

Tolerances can also be measured after acclimation, at 23°C for 24 hr in the case of heat tolerance and at 4°C for 24 hr in the case of cold tolerance. These measurements, together with measurements before acclimation, provide estimates of physiological flexibility (Table 3-1). Obviously, moreover, each of these traits can be considered as either individual or population traits.

Egg production per female is measured by isolating ovigerous females, each in one droplet of water, under a binocular microscope and counting the eggs in the egg sac. Counts are quite accurate up to 50 eggs.

Development time so far has been equated to age at first reproduction and measured only in females (age at first production of eggs). For example, a female becoming ovigerous and removed after t days has a development time of t days and cannot be confused with females becoming ovigerous later.

Viability and secondary sex ratio are estimated from counts of males and females in each brood at maturity, with the number of eggs produced known.

Analyses

Phenotypic variance among organisms classified into families can be partitioned as follows:

$$\sigma_P^2 = \sigma_A^2 + \sigma_{NA}^2 + \sigma_{Eg}^2 + \sigma_{Es}^2$$

where

σ_A^2 = additive genetic variance

σ_{NA}^2 = nonadditive genetic variance (dominance and epistasis)

σ_{Eg}^2 = general environmental variance (between individuals or families)

σ_{Es}^2 = special environmental variance (within individuals)

The first of these (σ_A^2) has been used as a measure of population flexibility and the last (σ_{Es}^2) as a measure of individual flexibility. For a comprehensive discussion of these terms see Falconer (1960).

Estimates of additive genetic variance are generally based on covariances between relatives. Details are available in Falconer (1960) and elsewhere. The particular methods used in my laboratory for the trait temperature tolerance are based on full-sibling covariances, in the absence of dominance and maternal variance, described in Bradley (1978a).

The measurement of σ_{Es}^2 has been made in three ways. Direct estimates were obtained from test-retest correlations, as described in Falconer (1960), the assumption being that such correlations would be perfect without variation within individuals in temperature tolerance. An alternative measurement is by subtracting σ_A^2 from σ_p^2 assuming σ_{NA}^2 and σ_{Eg}^2 are small (Bradley 1978a). The third method of measuring physiological flexibility did not produce a useful estimate of σ_{Es}^2 but did indicate the degree of adjustment by an individual in response to temperature change and was more useful in measuring potential for physiological adaptation. Individuals were assayed for temperature tolerance before and after exposure to a higher temperature and the differential was taken as a measure of flexibility (Bradley 1978b). The remaining components of phenotypic variance appear to be small or absent (Bradley 1978a).

To measure genetic divergence between treatments, between areas in the field, or between seasons, the general approach has been to raise progeny from corresponding samples of animals, all under the same conditions, usually $15°C$. Again, in the case of temperature tolerance, it suffices to raise the F_1 and not F_2 progeny, because maternal effects seem to be absent (Bradley 1978a). Any divergence observed between treatments, areas, or seasons can then be assumed to be genetic. For seasonal variability, parents from one season are stored at 4°C for several months and then allowed to produce progeny at the same time as animals from another season, the progeny then being tested together.

Results

Before the various models themselves are considered, pertinent evidence from a variety of experiments is reviewed here. Some of the data have been published previously. The results include evidence of genetic and physiological variance in temperature tolerance in *Eurytemora*, evidence that *Eurytemora* populations respond to natural selection, and data on genotype-environment interactions, on seasonal variability in temperature tolerance, and on environmental and genetic diversity in egg production, viability, and sex ratio.

Physiological and Genetic Variance

The presence of considerable physiological and genetic variance in temperature tolerance in *Eurytemora affinis* has been demonstrated previously (Bradley 1978a,b, 1979). Physiological variance was greater among females, whether measured from test-retest correlations (Falconer 1960), from residual variances (subtracting all variances except σ_{Es}^2, shown in Methods) or by acclimation tests (assaying before and after exposure to higher temperature). Estimates of physiological variance using the first two methods are shown in Table 3-2 (adapted from Bradley 1979). In every instance physiological variance was greater in females than in males. Females also acclimated more than males when exposed to elevated temperatures for varying periods, such acclimation occurring in as little as 3 hr (Bradley 1978b).

Table 3-2. Estimates of Physiological Variance in Males and Females[a]

Method	Estimate			
	1	2	3	4
1. Test-retest correlation (n = 24 per sex)				
Female 38.4				
Male 14.0				
2. Subtraction of genetic variance[b]				
Female	9.5	22.2	52.8	27.5
Male	2.2	3.5	9.2	11.0

[a] Adapted from Bradley (1979).
[b] Estimate 1 was from half and full sibling analyses, estimates 2 and 3 from full sibling analyses, and estimate 4 from offspring-parent regression analysis (see Table 3-3).

Table 3-3. Heritabilities of temperature tolerance in *Eurytemora affinis* using three methods of estimation[a]

	Experiment 1		Full siblings Expt. 2	Full siblings Expt. 3	Offspring-parent regression Expt. 4
	Half siblings	Full siblings			
Heritabilities					
Female progeny	0.40 ± 0.18[b]	0.20 ± 0.09	0.11 ± 0.10	0.28 ± 0.18	0.11 ± 0.44
Male progeny	0.84 ± 0.35	0.79 ± 0.24	0.89 ± 0.45	0.78 ± 0.29	0.76 ± 0.26
No. of families	19	68	23	36	57
Phenotypic variances					
Female progeny		12.1	29.9	73.6	176.4
Male progeny		11.6	33.1	46.5	58.7
No. of families	19	68	23	36	57

[a] From Bradley (1978a, 1979).
[b] S.E. calculated according to Falconer (1960).

Estimates of additive genetic variance, expressed as fractions of phenotypic variance or heritabilities are shown in Table 3-3, as are the phenotypic variances. These estimates are from experiments including half siblings and full siblings. As can be seen from comparisons of the heritabilities, there is no evidence of directional dominance in temperature tolerance, which would be indicated by greater heritability estimates from full-sibling than from half-sibling covariances and also from a quadratic component in the offspring-parent regression estimate. There is also no evidence of any maternal or common environmental effects within broods (Bradley 1978a).

Note should be made again of the sexual dimorphism in the expression of physiological and additive genetic variance, more of the former in females and more of the latter in males. These variances are almost complementary in this case, at least where phenotypic variances in the two sexes are comparable, because the remaining variance components (nonadditive genetic variance and environmental variance between individuals) are small. Because sex does not seem to be chromosomally controlled, although sex ratio seems to be genetically controlled (McLaren 1976), an explanation for the sex dimorphism based on sex-linked polygenes seems unlikely.

Genetic Variance Between Treatments

Evidence that the observed additive genetic variance in temperature tolerance is usable in natural selection and could be so maintained is provided by two sets of data.

The first set of data was derived from progeny of field samples of *Eurytemora* collected around a power plant east of Baltimore Harbor. The heated discharge from the power plant created greater than normal spatial variation in temperature. Temperatures at each station are shown in Table 3-4 next to the station number. As can be seen, some differences exist between stations. The differences are presumed to be genetic because progeny were all raised at 15°C (see Methods). These differences are not consistent with the ambient temperatures of the parental animals. However, in males, where genetic variance is expressed more (Bradley 1978a), animals originating from stations affected by discharge heat (stations 2, 3, 4, and 5) had collectively higher tolerances than those from the intake station (station 1). Furthermore, again in males, animals from station 2 the immediate discharge area, had markedly higher tolerances than those from station 1. The pattern for female progeny was quite different, perhaps because females are not selected for tolerance itself but rather for flexibility, as suggested later.

A second source of evidence for differentiation by natural selection in mean tolerances of populations is data from progeny of samples of animals grown in five different regimes for approximately 2 years, or between 20 and 70 generations depending on temperature (Table 3-5). These data have been reported in Ketzner and Bradley (1982). Suffice it to say that there is clear evidence of selection for temperature tolerance, particularly in the regime changing at 1°C per day in a cycle between 10°C and 22°C.

Whether genetic variance in temperature tolerance is actually translated into a response to selection seems to depend on the rate of change in the case of female progeny and on the magnitude of change in the case of males (Table 3-6). Note also that overall selection for temperature tolerance is significant in the males, but not in females.

Table 3-4. Mean temperature tolerances of *Eurytemora affinis* progeny from five field stations (raised at 15°C and tested before and after acclimation at 23°C for 24 hr)

	Before acclimation		After acclimation	
	Female[a]	Male	Female	Male
Station (ambient temperature)				
1 (20.1°C)	14.8	10.1	17.3	13.4
2 (23.7°C)	13.8	13.6	18.6	16.5
3 (22.9°C)	14.6	12.6	17.7	13.4
4 (20.6°C)	17.3	14.6	20.6	18.8
5 (23.6°C)	14.3	11.3	17.0	16.8
Overall mean	15.0	12.4	18.2	15.8
Contrasts (from ANOVA)				
1 vs. 2, 3, 4, 5		b		b
5 vs. 2, 3, 4		b		
4 vs. 2, 3	b		c	
3 vs. 2				b

[a] 20 animals per sex per station.
[b] p < 0.01.
[c] p < 0.05.

Table 3-5. Mean temperature tolerances (minutes) in progeny from five regimes, tested before and after acclimation at 23°C for 24 hr[a]

	10°C[c]	23°C[c]	15°C[c]	SW[c]	LW[c]
Preacclimation[b]	18.0	21.0	20.2	25.1	22.2
Postacclimation[b]	24.4	28.2	27.3	29.6	29.1

Contrasts (from ANOVA)	Preacclimation	Postacclimation
10°C, 15°C vs. 23°C	d	d
10°C vs. 15°C	d	e
SW, LW vs. 15°C	e	d
SW vs. LW	e	N.S.

[a] Adapted from Ketzner and Bradley (1982).
[b] 60 animals per mean (males and females combined).
[c] Regimes: 10°C, 23°C, 15°C, constant temperatures; SW, cycling at 1°/day between 10 and 22°C; LW, cycling at 1°/week between 10 and 22°C.
[d] p < 0.05.
[e] p < 0.01.

Table 3-6. Rate and magnitude of change imposed on parents and reflected in tolerances of progeny[a]

Range:	23°C			⟶	33°C	23°C ⟶ 38°C	Control
	Immed.	Immed.					
Rate:	+ 5 min	+ 10 min					
	at 33°C	at 33°C	1°C/5 min	1°C/30 min	1°C/30 min	15°C
Experiment:	A	B	C	D	E	F
Females:	16.9	23.5	19.4	15.4	19.3	19.7
Males:	16.8	19.5	18.3	16.0	20.5	15.4

Contrasts (ANOVA)	Females	Males
A, B, C, D, E vs. F (selection)	N.S.	b
A, B vs. C, D, E (rate of change)	c	N.S.
C vs. D (rate of change)	b	N.S.
C, D vs. E (magnitude of change)	N.S.	b
A vs. B (length of exposure)	b	b

[a] Adapted from Ketzner and Bradley (1982).
[b] $p < 0.01$.
[c] $p < 0.05$.

Genotype-Environment Interaction

Any investigation of seasonal adaptation must include the question of genotype-environment interaction. Three types of evidence are available so far. The relative responses of broods to heat and to cold are similar, or at least not negatively related, based on the results in Table 3-7. Correlations between hot and cold tolerances in individuals were positive, as were correlations between brood means. Second, in an experiment where broods were split, correlations between halves of broods grown in two salinities (5 and 12 ppt) were all at least 0.4. Brood X salinity interaction was not significant. Finally, rank correlations between preacclimation and postacclimation tolerances of broods were all positive and brood X acclimation interaction was significant in only one of ten analyses.

Seasonal Variability

More direct evidence on seasonal adaptation is shown in Table 3-8. Whereas tolerance clearly varies seasonally, the variability between seasons is not genetic, because postacclimation regressions were lower and progeny regressions not significant. It should be noted that collection dates producing progeny were rare in the summer months because of low frequency of adults, or of *Eurytemora* in general. When progeny were grown from parents held at 4°C, allowing progeny from separated months to be tested together, no clear trends were established. There was no consistent pattern in rankings of months, either from year to year or even between males and females in the same year. Hence the genetic diversity between seasons, if it exists, is small relative to total diversity.

Table 3-7. Relationships between hot and cold tolerances

	Males	Females
Individuals		
Correlations	0.32	0.54
	N = 24	N = 24
Other estimates lower, but never negative		
In broods (means)		
Rank correlations	0.512 ± 0.092	0.118 ± 0.092
Product-moment	0.126	0.108
correlations	N = 60	N = 57

Diversity in Life History Traits

Data have been collected on variation in egg production between wild populations and between the laboratory regimes mentioned earlier and on viability and sex ratio between laboratory regimes. According to Table 3-9, egg production decreased significantly with increased temperature.

Egg production was also lower in the regimes with higher temperatures (Table 3-10), especially at 23°C. The divergence between regimes was not genetic, however, as can be seen from the progeny means (Table 3-10). A similar result in viability differences between regimes is shown in Table 3-11, with average viability lowest in broods from the 23°C regime and possibly lower in the broods of female progeny from the 23°C regime. The number of broods providing the data from the 23°C regime was considerably lower. Sex ratios, expressed as number of males adjusted for total brood size, were also lower in the 23°C regime. Again the difference was not genetic (Table 3-12).

Some preliminary data on relationships between temperature tolerance and other fitness traits have been collected on 51 broods. In those data low (positive) correlations were observed between egg number, adult number, and viability in a brood, on the one hand, and average temperature tolerances of the progeny in the brood (all around 0.2) on the other. The only indication of any negative correlations were between egg number and viability (−0.4) and temperature tolerance and number of adult females (−0.2). In a second set of data, tolerances of 176 female progeny were negatively correlated (−0.1) with age at first reproduction (egg to egg development time in females), and with viability of the original brood (−0.1). As in the first set of data, viability and egg

Table 3-8. Linear regressions (±SE) of temperature tolerance (minutes) on ambient temperature (°C)

	Males	Females
Wild-Caught[a]		
Preacclimation	1.22 ± 0.11	1.12 ± 0.12
Postacclimation	0.47 ± 0.10	0.48 ± 0.10
Progeny (raised at 15°C)	0.18 ± 0.10	0.14 ± 0.11

[a] Regressions based on 24, 20, and 16 collection dates.

Table 3-9. Egg production per female at different ambient temperatures[a]

Temperature range	7.5-24°C
Overall average egg production	64.8 ± 21.3
Regression of egg production on temperature	-1.17 ± 0.44

[a] Based on 32 sets of observations over 2 years.

number in the brood were negatively related (-0.3 in this case). This latter correlation may be a result of the laboratory rearing methods used. Useful data could not be collected on the males in this second set because they had to remain in the flasks until all the females had produced egg sacs and hence were generally beyond the age at which tolerance is measured.

Comparison of Models

In light of the data discussed above a number of general and specific models can be examined. The general models are shown in Table 3-13 and are addressed to the question of how *Eurytemora* adapts to seasonal temperatures. The specific models are intended to explain how genetic variance in temperature tolerance in *Eurytemora* is maintained, in the presence of considerable physiological variance. I am assuming that the genetic variance observed in the laboratory is also expressed in the natural environment, particularly because genotype-environment interaction is small or absent.

If a particular genotype were adapted to the whole range of temperatures genetic variance would be expected to disappear because of random drift, unless the heterozygote were more homeostatic (Gillespie 1974b). I have found no evidence that

Table 3-10. Diversity among regimes[a] in egg production in parents collected from the regimes and in progeny raised at 15°C

	Regime				
	10°C	15°C	23°C	SW	LW
Parents	33.1	33.6	11.1	24.3	24.9
(No. of broods)	(205)	(237)	(32)	(208)	(300)
Progeny	24.4	24.7	23.6	25.1	25.2
(No. of broods)	(62)	(62)	(56)	(60)	(60)

Parents
  SW, LW, 23°C < 10°C, 15°C (p < 0.05)
  10°C vs. 15°C   N.S.
  SW, LW > 23°C (p < 0.05)
  SW vs. LW   N.S.

Progeny
  No significant contrasts

[a] See under Table 3-5.

Table 3-11. Diversity among regimes in egg to adult viabilities in broods (raised at 15°C) from females sampled from five regimes, and from F_1 progeny also raised at 15°C

	Regime				
	10°C	15°C	23°C	SW	LW
Parental broods	50.9	42.8	27.2	44.3	55.1
(including broods not	(29.5)	(24.4)	(10.1)	(32.8)	(20.9)
maturing)					
No. of broods	118	136	12	153	115
Progeny broods	30.9	36.1	30.0	33.5	37.5
(including broods not	(20.4)	(27.4)	(16.6)	(24.4)	(25.7)
maturing)					
No. of broods	41	47	31	39	46

Parental broods (transformed data)
  SW, LW, 23°C vs. 10°C, 15°C N.S.
  10°C vs. 15°C N.S.
  SW, LW > 23°C (p < 0.05)
  LW > SW (p < 0.01)

Progeny broods
  15°C > 10°C (p < 0.01)
  Others N.S.

Table 3-12. Diversity among regimes in sex ratio (percentage males) in broods (raised at 15°C) from females sampled from five regimes and from F_1 progeny also raised at 15°C

	Regime				
	10°C	15°C	23°C	SW	LW
Parental broods	32.9	39.4	33.3	48.7	43.2
(No.)	(118)	(136)	(12)	(153)	(115)
Progeny broods	39.1	48.1	43.2	46.9	42.7
(No.)	(41)	(47)	(31)	(39)	(46)

Parental broods (No. of males adjusted for total brood size)
  SW, LW, 23°C vs. 10°C, 15°C N.S.
  10°C vs. 15°C N.S.
  SW, LW > 23°C (p < 0.05)
  SW vs. LW N.S.

Progeny broods
  No significant contrasts

Table 3-13. Alternative hypotheses for adaptation of *Eurytemora affinis* to seasonal temperatures

---

Physiological
    Individual genotypes adapted to range
    Little genetic variance

Genetic
    Population adapted to entire range
    Genetic variance maintained

Physiological and genetic
    Each genotype has a range of tolerance
    Intermediate levels of genetic variance

---

physiological flexibility is maximum at intermediate tolerance levels, or the levels at which heterozygotes would be with largely additive genetic variance. Hence the purely physiological model is unlikely (Table 3-13).

The purely genetic model is also unlikely unless the physiological variance has no adaptive significance. Therefore, any model proposed must consider both genetic and physiological variance.

Maintenance of Genetic Variance

Given that there is both genetic and physiological variance, expressed to differing degrees in the two sexes (Tables 3-2 and 3-3), the question of adaptation can be redefined.

Because selection is effective in differentiating between populations in the natural environment, at least when spatially influenced by a power plant (Table 3-4), and in the laboratory (Table 3-5), selection may well be responsible for maintaining the observed genetic variance. Assuming that a random drift (neutral) model is unlikely in the case of a character as important as temperature tolerance (closely related to survival at temperatures around $30°C$, as stated in Methods), the question then becomes which kind of selection is most plausible.

There are basically three selection models, as summarized in Table 3-14, although the variable selection model has a number of possible forms. Of all the possible models, two are chosen as most likely, given the data available.

Heterozygote Superiority

Any model requiring net dominance in the temperature tolerance character is not likely because no evidence of dominance variance has been found. Estimates of heritability from full and half siblings are comparable, although variable (Table 3-3), as are estimates from full siblings and from offspring-parent regressions. Furthermore, as indicated in Bradley (1978a), there is no deviation from linearity in the offspring-parent regressions. A variant of this model by Gillespie (1974) where the superiority of

Table 3-14. Alternative hypotheses for maintenance of genetic variance in temperature tolerance in *Eurytemora affinis*

Hypothesis	Plausibility
Heterozygote superiority	Unlikely: Directional dominance variance small or absent
Variable selection	
Cyclical selection	Unlikely: Hot and cold tolerance not opposite traits; slight evidence for seasonal genetic trends in temperature tolerance; no evidence of GE interaction
Frequency-dependent selection	Unlikely in polygenic trait
Density-dependent selection	Unlikely
Disruptive selection	
Temporal	Unlikely: see Cyclical selection
Spatial	Unlikely: Selection in same direction in patches; fine-grained environment
Between sexes	Possible: If selection for tolerance in males, for flexibility in females, and r_g negative
Genetic homeostasis	Possible: If tolerance and other fitness traits negatively related

heterozygous genotypes is a result of their greater homeostasis, has been discussed earlier and is also considered unlikely.

Heterozygous genotypes may be optimum even without dominance and in such cases selection is said to be stabilizing. Stabilizing selection cannot maintain genetic variance indefinitely (Robertson 1956), although Lande (1976) has shown using experimental data from *Drosophila*, maize, and mice that genetic variance can be maintained by mutation in polygenic characters, even with strong stabilizing selection. The possibility that variation in temperature tolerance is being continuously generated by mutation cannot be ruled out.

Cyclical Selection

Cyclical or seasonal selection is an obvious possibility because temperatures vary over a wide range between seasons, as much as 30°C in the Chesapeake Bay. However, little evidence, direct or indirect, of seasonal selection for temperature tolerance has been found. No consistent pattern has been observed in tolerances of progeny from different seasons (Table 3-8), or from contemporary comparisons of progeny from parents collected at different seasons and stored at 4°C (see Methods). It is possible that the differences are below the levels that presently can be detected and also that the rate of change in temperature is not sufficient to induce dramatic changes in gene frequency (Table 3-6).

Indirect evidence on seasonal or cyclical selection comes from a series of experiments on genotype-environment interactions, the extreme form of which considers temperature tolerance in general to be the character and examines whether this general tolerance character is expressed differentially by genotype in high and in low temperatures. The results of these experiments on hot and cold tolerances have been described earlier and shown in Table 3-7. They do not suggest the necessary interaction. At most, the expressions of the character in high and low temperatures are unrelated. Other results concerning genotype by environment interaction were described, in these cases considering heat tolerance as the character. No evidence of such an interaction was obtained. Furthermore, in view of the dimorphism between the sexes in expression of genetic differences in temperature tolerance, it is important to recall that there was no genotype X sex (environment) interaction in temperature tolerance, when the means of the two sexes were compared by brood (Bradley 1978a).

Frequency-Dependent and Density-Dependent Selection

Selection varying by genotype frequency or by density is not easily disproved. The former seems unlikely, at least in the form reported by Kojima and Yarbrough (1967), if there are more than a few genes segregating for the character. Moreover, because the greatest selection pressure, at least for tolerance to high temperatures, occurs at times of lowest density and when reproductive rates, at least per individual, are lowest, the usual density-dependent models of natural selection do not seem to apply.

Disruptive Selection

Three forms of disruptive selection are listed in Table 3-14. To this could be added differential selection at different life stages. The basic models for the maintenance of genetic variance in temporal and spatial heterogeneity were given by Haldane and Jayakar (1963) and by Levene (1953). These models recently have been reviewed and combined into models with both temporal and spatial components by Ewing (1979). They have also been extended by a number of authors whose work has been reviewed by Hedrick et al. (1976) and by Roughgarden (1979), for example. The critique by Maynard Smith and Hoekstra (1980) also should be noted.

A feature of all of these temporal and spatial models is selection favoring different genotypes in different environments in space or time. There is no evidence for such differential selection in the case of temperature tolerance in *Eurytemora*, as can be seen from the lack of genotype-environment interaction, indicating that heat tolerance is the same trait in every environment. So without further consideration of these widely investigated models, I wish to propose two possible models, one under the category of disruptive selection. Both of these models date back to the 1950s.

Mather (1955) suggested that a form of disruptive selection between the sexes could lead to a balanced polymorphism. In the case of *Eurytemora* this would most likely mean selection for physiological flexibility in females and for temperature tolerance itself in males (Ketzner 1979). In order for the model to be sufficient there should be a negative genetic correlation between flexibility and tolerance, such as would occur if they were selected together (Falconer 1960), and the sexual dimorphism itself should not have genetic variability (Russell Lande, personal communication). I do not have information on the genetic correlation between tolerance and flexibility. There is significant genetic variance in the flexibility trait (Ketzner 1979), and there is some evidence that progeny with lower average tolerances among the regimes have higher average flexibilities (Table 3-5), suggesting selection in opposite directions. Also, selection for tolerance seems to occur differently in males and females (Table 3-6). If selection in males and females were in fact for different traits, genetic variance in both would be maintained because of the 50% (maximum) gene flow between the sexes each generation.

Genetic Homeostasis

The last model, genetic homeostasis, also depends on sexual reproduction (Lerner 1954). Lerner proposed that selection for some desired quantitative trait is resisted by natural selection, because the optimum genotypes for reproductive fitness are heterozygous. As with the disruptive selection model, this model requires genetic variance in temperature tolerance in at least some life history traits and negative genetic correlations between them, whether because of pleiotropy or tight linkage. In *Eurytemora*, selection for temperature tolerance during the warm summer months would move populations away from gene frequency equilibria which would be restored upon relaxation of selection during the remainder of the year.

There is clearly an influence of temperature on egg production (Table 3-9), but no evidence of genetic divergence among the regimes mentioned earlier (Table 3-10). There is some evidence, not convincing, of genetic divergence in viability (Table 3-11). The numbers are small and maternal (brood) influences on viability have not been ruled out. There is also no evidence for genetic divergence in sex ratio (Table 3-12).

Although the importance of egg production and viability is clear, the role of sex ratio is not. Heinle (1970) has suggested that an excess of males in *Eurytemora* should be advantageous, given the need for copulation to produce each egg sac.

At present two components of the genetic homeostasis model seem reasonable. There is certainly genetic variance in temperature tolerance, and McLaren (1976) has reported significant genetic variance in age at maturity and in body size but not in mortality or sex ratio in *Eurytemora herdmani*. The third requirement, a negative genetic relationship between temperature tolerance and life history traits, has not been demonstrated, as was indicated at the end of the Results section. There were slight negative correlations between female tolerance and development time (egg to egg or equivalent to age at first reproduction) and between female tolerance and viability of the brood. At the environmental (regime) level, mean temperature tolerance seems to be negatively related to egg production (Tables 3-3 and 3-10), but this relationship disappears in the progeny raised at 15°C.

Combined Model

Aside from Lande's mutation model mentioned earlier, the only plausible models appear to be the disruptive selection model and the genetic homeostasis model. One hesitates to suggest a further model; however, the latter two can be combined into a single descriptive model, as follows.

First, the various life history traits are combined into one parameter, production (P). The other traits in the combined model are temperature tolerance (A) and physiological flexibility (F), defined as the difference between post acclimation and preacclimation tolerance in Table 3-1. Including the two sexes, there are six variables in the model. It has been shown already (Bradley 1978a) that temperature tolerance is the same trait genetically in males and in females. Assuming for the moment that this is true for other traits, then the combined model would have three variables.

In order for an equilibrium to exist, two of the variables should be positively related, for example P and F, and the other pairs of variables (A and F and A and P) should be negatively related. These latter negative correlations are requirements of the separate models, namely disruptive selection and genetic homeostasis. A positive relationship between flexibility and the life history traits is not unreasonable. As already stated, the data are inconclusive but the definitive experiments can be done.

If either the F or P trait is shown to be differently expressed, on average, between males and females then the model includes sex variables and 15, not 3 correlations. Furthermore, the P trait obviously includes several other traits, adding to the complexity. Whatever the required number of variables needed to explain the system adequately, the correlations should persist, suggesting that the ultimate question may concern the maintenance of genetic correlations as well as genetic variance.

Acknowledgments. I am grateful to Beatrice Boffen, Diana Foster, Stephen Harrison, Kenneth Keeling, Phyllis Ketzner, and Thomas Williams who assisted in one way or another with this project. The work was supported by NSF Grants DEB 77-26921 and DEB 80-10903 and also by Contract No. P39-78-04 from the Power Plant Siting Program, State of Mayland.

References

Battaglia, B.: Genetic aspects of benthic ecology in brackish waters. In: Estuaries. Lauff, G. W. (ed.). Washington, D.C.: AAAS Publ. No. 83, 1967, pp. 574-577.

Bradley, B. P.: The anomalous influence of salinity on temperature tolerances of summer and winter populations of the copepod *Eurytemora affinis*. Biol. Bull. 148, 26-34 (1975).

Bradley, B. P.: The measurement of temperature tolerance: Verification of an index. Limnol. Oceanog. 21, 596-599 (1976).

Bradley, B. P.: Increase in range of temperature tolerance by acclimation in the copepod *Eurytemora affinis*. Biol. Bull. 154, 177-187 (1978 a).

Bradley, B. P.: Genetic and physiological adaptation of the copepod *Eurytemora affinis* to seasonal temperatures. Genetics 90, 193-205 (1978 b)

Bradley, B. P.: Genetic and physiological flexibility of a calanoid copepod in thermal stress. In: Energy and Environmental Stress in Aquatic Systems (J. H. Thorp, J. W. Gibbons, Eds.), Washington, D.C.: Dept. of Energy CONF-771114, 1979.

Engel, R. A.: *Eurytemora affinis*, a calanoid copepod new to Lake Erie. Ohio J. Sci. 62, 252-255 (1962).

Ewing, E. P.: Genetic variation in a heterogeneous environment VII. Temporal and spatial heterogeneity in infinite populations. Am. Nat. 114, 199-212 (1979).

Falconer, D. S.: Introduction to Quantitative Genetics. Edinburgh: Oliver and Boyd, 1960.

Gillespie, J. H.: Polymorphism in patchy environments. Am. Nat. 108, 145-151 (1974 a).

Gillespie, J. H.: The role of environmental grain in the maintenance of genetic variation. Am. Nat. 108, 831-836 (1974 b).

Gillespie, J. H.: A general model to account for enzyme variation in natural populations. Am. Nat. 110, 809-821 (1976).

Gonzalez, J.: Critical thermal maxima and upper lethal temperatures for the calanoid copepods *Acartia tonsa* and *A. clausi*. Mar. Biol. 27, 219-223 (1974).

Haldane, J. B. S., Jayakar, S. D.: Polymorphism due to selection of varying direction. J. Genet. 58, 237-242 (1963).

Hedrick, P. W., Ginevan, M. E., Ewing, E. P.: Genetic polymorphism in heterogeneous environments. Ann. Rev. Ecol. Syst. 7, 1-32 (1976).

Heinle, D. R.: Population dynamics of exploited cultures of calanoid copepods. Helgol. Wiss. Meere. 20, 360-366 (1970).

Ketzner, P. A.: The effect of constant and varying temperature regimes on the genetic and physiological flexibility of the copepod *Eurytemora affinis*. Unpubl. M.S. Thesis, Dept. of Biological Sciences, University of Maryland, Baltimore, 1979.

Ketzner, P. A., Bradley, B. P.: Rate of environmental change and adaptation in the copepod *Eurytemora affinis*. Evolution (1982).

Kojima, K., Yarbrough, K.: Frequency-dependent selection at the esterese-6 locus in *D. melanogaster*. Proc. Natl. Acad. Sci. (US) 57, 645-649 (1967).

Lande, R.: The maintenance of genetic variability by mutation in polygenic characters with linked loci. Genet. Res. Cambr. 26, 221-235 (1976).

Lerner, M.: Genetic Homeostasis. New York: John Wiley and Sons, 1954.

Levene, H.: Genetic equilibrium when more than one ecological niche is available. Am. Nat. 87, 331-333 (1953).

Mather, K.: Polymorphism as an outcome of disruptive selection. Evolution 9, 52-61 (1955).

Maynard Smith, J., Hoekstra, R.: Polymorphism in a varied environment: How robust are the models? Genet. Res. Cambr. 35, 45-57 (1980).

McLaren, I. A.: Inheritance of demographic and production parameters in the marine copepod *Eurytemora herdmani*. Biol. Bull. 151, 200-213 (1976).

Robertson, A.: The effect of selection against extreme deviants based on deviation or on homozygosis. J. Genet. 54, 236-249 (1956).

Roughgarden, J.: Theory of Population Genetics and Evolutionary Ecology. An Introduction. New York: Macmillan, 1979.

Slobodkin, L. B., Rapoport, A.: An optimal strategy for evolution. Q. Rev. Biol. 49, 181-199 (1974).

Chapter 4

Evolution of Seasonal Adaptations and Life History Traits in *Chrysopa:* Response to Diverse Selective Pressures

CATHERINE A. TAUBER and MAURICE J. TAUBER*

Introduction

Earth's seasonal cycles have multiple direct and indirect influences on the life histories of organisms. Growth, development, and reproduction must be synchronized with the seasonal presence of energy resources, mates, and favorable physical conditions. Also, the seasonal occurrence of vulnerable stages must be timed to avoid periods of physical extremes and periods when parasites, predators, and competitors occur at high densities. In a seasonally variable environment, therefore, growth, development, and reproduction are interrupted by periods of dormancy and migration; these periods subserve survival, and they maximize the probability of growth, development, and reproduction occurring at an appropriate time and place.

Life history traits (schedules of fecundity and mortality) and seasonal adaptations are intertwined so that selection pressure on one causes, to varying degrees, selection pressure on the other (Masaki 1978). When considering the evolution and expression of life histories, it is therefore necessary to consider the ecophysiology and genetics of seasonal traits. Among insects, for example, diapause is a major means of achieving seasonal synchronization. This seasonal adaptation, however, not only affects dormancy and survival during adverse conditions, but it also influences the timing and rates of growth, development, and reproduction during favorable periods.

The physical extremes of a locale set the ultimate constraints on the seasonal timing and rate of growth and reproduction of a population. Within these limits, interspecific interactions (e.g., occurrence of hosts, natural enemies, and competitors) and intraspecific interactions (e.g., occurrence of mates) mold a species' life history pattern. These environmental patterns, the physical and biotic, interact with the genetic capability of organisms to determine the fitness of populations (e.g., Dingle 1974, Dingle et al. 1977, Istock 1978, Istock et al. 1976), the evolution and maintanance of polymorphisms (e.g., Bradshaw 1973, Vepsäläinen 1978), and, in some cases, speciation and the maintenance of reproductive isolation (e.g., Masaki 1978, Tauber & Tauber 1977a, b, 1981b).

*Department of Entomology, Cornell University, Ithaca, New York 14853 U.S.A.

One method for elucidating the evolution of life history patterns is to analyze the relationship between seasonal adaptations and life history traits as it occurs within and among geographical populations. We take this approach in this chapter. We begin by briefly reviewing the types of intra- and interpopulation variability in insect seasonal cycles and their likely effects on the expression of life history traits. Next, we discuss examples from our work with *Chrysopa* that illustrate several types of seasonal and life history variation. Finally, we examine the geographical variability in the interaction between seasonal and life history traits and the various selective forces that may mold the evolution of the interaction.

Variability in Seasonal Adaptations and Life History Traits in Insects

For this discussion, variation in seasonal adaptations and life history traits can be divided into two types: continuous and disjunct. Both types of variation can occur within and among populations, and both can have a geographical component.

Continuous Variation

Continuous variation in life history or seasonal adaptations involves quantitative characteristics that vary around a mean value. Among insects the most commonly studied are the rates of nondiapause and postdiapause growth and development, fecundity, adult size, critical photoperiod for diapause induction, diapause depth (or duration), temperature requirements for rapid diapause termination, and duration of the period of sensitivity to diapause-inducing stimuli (see Danilevskii 1965, Dingle 1974, 1978, Lees 1955, Masaki 1961, 1978, Tauber & Tauber, 1978, 1981b).

The critical photoperiod for diapause induction and the depth of diapause have been shown to be polygenic characters that are subject to relatively rapid alteration by artificial selection. Typically, these traits have high heritabilities (high additive genetic variance) (Dingle 1974, Dingle et al. 1977). As a consequence of differences in selection, such continuous variation in quantitative traits may contribute in large measure (although not exclusively) to the evolution of smooth and stepped clines (e.g., Ando 1979, Masaki 1978), the extension of geographical distribution and host ranges (Bush 1974, Krysan et al. 1977, Sims 1980), and allopatric and parapatric speciation (Masaki 1978, Tauber & Tauber 1981b). It may also play a role in post-speciation divergence of sympatrically and allopatrically derived species (see Tauber & Tauber 1977b, 1981b).

Disjunct Variation

Life history and seasonal traits also can vary in a disjunct or qualitative fashion. Examples among insects include polymorphic emergence time (Bradshaw 1973, Waldbauer 1978), differences in diapausing stages (Masaki 1978), univoltine versus multivoltine life cycles (Danilevskii 1965, Tauber & Tauber 1978), presence or absence of estival diapause (Tauber & Tauber 1973c, 1978, 1981a), and alternate morphological forms associated with alternate seasonal and reproductive cycles (Blackman 1971, 1974, Bradshaw 1973, Denno & Grissell 1979, Vepsäläinen 1974, 1978,).

In some cases, disjunct variation serves as an adaptation to heterogeneous or fluctuating environments (see Stearns 1976), and the degree to which insect species express polymorphic life history and seasonal adaptations often serves as a measure of the unpredictability of their habitat (Istock 1978, Southwood 1977, Stearns 1976, Vepsäläinen 1978). Disjunct variation also plays an important role in the evolution and maintenance of reproductive isolation between sympatric species (Tauber & Tauber 1976a, 1977a, b).

Diverse genetic mechanisms—single genes, "supergenes," and polygenes—underlie disjunct variation in life history and seasonal adaptations. In polymorphic populations, these systems function as genetic "switch" mechanisms, genetic "rheostats," or "threshold" mechanisms (Dingle et al. 1977, Lumme & Oikarinen 1977, Vepsäläinen 1978). In reproductively isolated populations they serve as the genetic bases for asynchronous life cycles (Tauber et al, 1977).

In most examples, life history and seasonal traits form coordinated sets of characters that vary in an interrelated manner. Coordination of traits has been demonstrated to underlie geographical adaptations (Goldschmidt 1934, Kidokoro & Masaki 1978, Masaki 1978, 1979) and also to subserve a polymorphic life cycle (Denno & Grissell 1979, Vepsäläinen 1978). In the next sections we provide additional examples of coordinated traits in polymorphic and geographically differentiated populations from *Chrysopa*, and we also give an example where coordination of life history traits functions to maintain seasonal isolation between potentially interbreeding, sympatric species.

Life History Variability in *Chrysopa*

The *Chrysopa carnea* species-complex (Neuroptera: Chrysopidae), the common green lacewings, occurs throughout the Holarctic region, where it is typified by diverse species and strains. In North America, and perhaps in the Old World as well (Alrouechdi & Canard 1979), these species and strains exhibit considerable variation in their seasonal and life history traits. Much, but not all, of the variation in life history traits appears to be related to variation in local seasonal conditions, and so life history and seasonal adaptations are considered together here.

The species and strains in this species-complex lay their eggs at the end of silken stalks. The larvae are predaceous, feeding on a variety of soft-bodied arthropods. Larvae undergo three moults; the mature third-instar larva spins a silken cocoon within which the final moult, to the pupa, takes place. With the exception of one strain, adults of the *Chrysopa carnea* species-complex are not predaceous; they feed on honeydew and pollen. Diapause occurs only in the adult stage, and overwintering and estivation (when they occur) are accomplished by adults.

Diapause in the species-complex is characterized by the lack of ovarian development, the lack of sperm transfer to the seminal vesicles (MacLeod 1967), accumulation of fat body, lack of reproductive behavior, and in some cases the assumption of autumnal (brownish) coloration (Tauber & Tauber 1973a, Tauber et al. 1970).

In North American populations of the *C. carnea* species-complex, the two major types of variation in life cycles, continuous and disjunct, are recognized. Examples of each, beginning with the simplest—the *carnea*-type life cycle—are described in the following sections.

The *Carnea* Type (Long-Day Reproduction)

The *carnea* category is characterized by a multivoltine life cycle and the lack of an estival dormancy (diapause). Reproduction occurs under long day lengths, throughout spring and summer, resulting, for example, in three generations per year in the Ithaca, New York area. At the end of summer, when day lengths fall below a critical level (critical photoperiod), adults enter reproductive diapause. Diapause persists in response to the actual duration of the short day lengths of autumn and early winter (Tauber & Tauber 1973b). No particular stimulus terminates diapause, and the timing of diapause termination shows considerable intrapopulation variation. In the Ithaca population diapause ends between the end of January and the middle of March.

After diapause ends, temperature determines the rate of postdiapause reproductive development. In the Ithaca population, postdiapause oviposition starts after an accumulation of 100 heat-degree days above a threshold of 4°C (about the end of April) (Tauber & Tauber 1973a). The initiation of oviposition can occur without the female receiving an external source of protein, but sustained oviposition requires proteinaceous food. In nature, aphid honeydew and pollen are the primary sources of the required nutrients (Hagen & Tassan 1966, Sheldon & MacLeod 1971).

Under conditions of adequate food, the rates of preimaginal growth and development are dependent on temperature. In the Ithaca population, about 376 heat-degree days above 9.5°C are required to complete preimaginal development (Tauber & Tauber 1973a).

Continous Variation

Several characteristics of the *carnea*-type life cycle show continuous, genetic variability. This variability adapts populations to local conditions without altering the basic, long-day reproduction and multivoltine life cycle.

Critical Photoperiod. The critical photoperiod for diapause induction varies greatly with locality (Tauber & Tauber 1972). In general, there is a north-south gradient (from long to short) in the length of the critical photoperiod (Table 4-1). In the northern hemisphere, winter comes early and late-summer day lengths are long in northern localities; under these conditions a long critical photoperiod subserves the appropriate timing of the preparatory physiological and behavioral changes associated with diapause induction.

Table 4-1. Geographical variation in *Chrysopa carnea* critical photoperiod and diapause depth (T = 24 ± 1°C)

	Latitude (∿°N)	Critical photoperiod L:D[a]	Diapause duration (X̄ ± S. D. days at L:D 10:14)
Portage la Prairie, Manitoba, Canada	49.5	14.5:9.5	143 ± 14
Brownstown, Washington	46.5	14.5:9.5	138 ± 25
Ithaca, New York	42.5	13.5:10.5	87 ± 8
Corcoran, California	36	14:10	96 ± 13
Sedona, Arizona	35	12.5:11.5	45 ± 14
Chandler, Arizona	33.5	12.5:11.5	29 ± 13
Brawley, California	33	12.5:11.5	31 ± 12
Quincy, Florida	30.5	13:11	92 ± 20
El Sueco Junction, Chihuahua, Mexico	30	12:12	42 ± 13

[a] L:D = Light:Dark

Duration of Diapause. The duration (or depth) of diapause is in part, intrinsically determined, and this character shows geographical variation among populations, usually in a direct relationship with the critical photoperiod (Table 4-1; Figure 4-1). However, this trait is not considered of primary importance in the *carnea*-type life cycle because diapause ends spontaneously in midwinter under natural conditions (Tauber & Tauber 1973b).

In many species the physiological state of diapause is largely complete by early to midwinter, even though dormancy persists until mid- to late spring (Tauber & Tauber 1976b). The duration of diapause, provided it is not so short as to allow premature development during warm autumnal periods, probably is not subject to strong natural selection. Because of this, diapause duration often shows considerable intrapopulation variation, and it may not be a life history trait of major importance to survival or fitness. However, the use of this character in the analysis of life history evolution underscores the need for verifying its specific function in nature.

We have selected for both shortened diapause and increased response to diapause-inducing photoperiods in the adult stage. Our selection regimens reduced the duration of diapause by 62 days in three generations (Table 4-2). Selection for increased adult response to diapause-inducing photoperiods resulted in an increase from 0 to 64% in the proportion of females that entered diapause without ovipositing. Simultaneously, the time to enter diapause in these females decreased from 64 to 11.5 days (Table 4-2). Therefore, both characters are amenable to rapid alteration by artificial selection.

Our results are similar to those that Dingle (1974) obtained from the milkweed bug, *Oncopeltus fasciatus*. They indicate high additive genetic variance in diapause duration and in the associated photoperiodic requirements for diapause induction.

Postdiapause Reproduction. The thermal requirements for the initiation of postdiapause reproduction show considerable geographical variation (Table 4-3). This char-

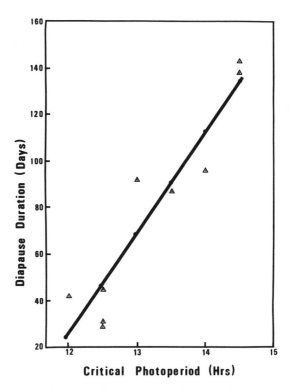

Figure 4-1. Correlation between the length of the critical photoperiod for diapause induction and diapause duration in *Chrysopa carnea*.

acter is probably very important in fine tuning the initiation of reproduction to the vernal conditions at diverse localities. Both the degree of predictability and the timing of the onset of the growing season may play a role in determing this character (Tauber & Tauber 1978).

With the exception of the most northern and most southern populations (Alaska and Mexico), there is an inverse relationship between latitude and the thermal threshold (t) for postdiapause development (Table 4-3). This general trend is associated with a geographical trend in the onset of the growing season. Low t values generally occur in northern populations that experience cool vernal conditions. High t values typify southern populations.

Preimaginal Growth. There is considerable variability in the thermal requirements for growth and development of preimaginal stages (Tables 4-4 and 4-5). We have analyzed the thermal requirements for development in six populations from various localities in North America. Each population was tested under a range of temperatures from 15.6 to 26.7°C. Using the linear regression method of analysis of the temperature-developmental rate relationship (see Obrycki & Tauber 1981), we determined the threshold (t) for development and the number of heat-degree days (K) required for preimaginal development of each population (Table 4-4). Threshold values ranged from 9.5 to 10.9°C and K values ranged from 300 to 378 heat-degree days above t.

Table 4-2. Selection for diapause duration and speed of diapause induction in *Chrysopa carnea*

	Selection for reduced duration of diapause[a]	
Generation	No. days from emergence to first oviposition	
1	69	
2	23	
3	17	
4	7	

	Selection for increased response to diapause-inducing day lengths by reproductively active females[b]	
Generation	No. of days of oviposition before diapause induction	% females entering diapause without initiating oviposition
1	64	0
2	60	4.2
3	56.4	52.5
4	11.5	64.5

[a] All individuals were reared and maintained under LD 10:14 (24 ± 1°C). Individuals (No. ≅ 20) from generations 2-4 originated from the first eggs laid by the previous generation.
[b] All individuals were reared under LD 16:8 and transferred to LD 10:14 on day of adult emergence. Each cage contained one pair of adults reared from eggs laid by the first four pairs of the previous generation to enter diapause. (No. cages each generation = 20-30; T = 24 ± 1°C.)

Our data illustrate that considerable variability in the thermal requirements for growth and development occurs among populations (Table 4-5). The *mohave* population from Strawberry Canyon, California, has a relatively high t value and a relatively large K value (Table 4-4). In comparing the developmental times of this population with the Alaskan population, which has a relatively low t value and a very low K value, it is seen that at 15.6°C the developmental rate of Strawberry Canyon immatures was faster than that of the Alaskan population (Table 4-5). However, the Strawberry

Table 4-3. Geographical variation in thermal requirements [heat-degree days (K) above threshold (t), °C] for postdiapause development of *Chrysopa carnea*

	Latitude (∼ °N)	t	K
Goldstream Valley, Alaska	65	9.4	43.3
Portage la Prairie, Manitoba, Canada	49.5	6.6	89.3
Brownstown, Washington	46.5	7.2	61.9
Ithaca, New York	42.5	8.6	80.5
Corcoran, California	36	9.2	67.3
Sedona, Arizona	35	9.1	69.1
Quincy, Florida	30.5	9.9	68.1
El Sueco Junction, Chihuahua, Mexico	30	5.1	81.6

Table 4-4. Thermal requirements [heat-degree days (K) above threshold (t), °C] for development (egg-to-adult emergence) of geographical populations of the *Chrysopa carnea* species-complex

	t	K
C. carnea, Goldstream Valley, Alaska	9.8 ± 1.3	300.4 ± 17.8
C. carnea, Ithaca, New York	9.5 ± 0.2	376.7 ± 8.2
C. carnea, El Sueco Junction, Chihuahua, Mexico	10.5 ± 0.3	333.3 ± 6.0
C. carnea, *mohave* strain, Strawberry Canyon, California	10.2 ± 1.2	358.3 ± 26.3
C. downesi, St. Ignatius, Montana	10.9 ± 0.2	355.9 ± 10.8
C. downesi, Ithaca, New York	10.8 ± 0.1	378.0 ± 10.2

Canyon population also suffered severe mortality (72%) at 15.6°C. Only those individuals that were able to complete development within a very short period of time were able to survive this low temperature. Under such selection pressure, a new adaptive mean might arise very rapidly.

Disjunct Variation

The continuous variation described above results in minor modifications of the life cycle; in addition, we found three examples of disjunct or major variations in the life cycles of the *C. carnea* species-complex. Two examples result in univoltine life cycles (one generation a year), and one results in a variable life cycle, having one to three generations a year.

The downesi *Type (Short-Day/Long-Day Reproduction).* This variation is typified by a univoltine life cycle with reproduction and development restricted to spring. In this case, the life cycle is primarily determined by photoperiodic influences on reproductive diapause; there is a short-day/long-day requirement for the initiation of oviposition.

Laboratory studies showed that constant day lengths of any duration result in diapause (Table 4-6A). Diapause was averted only if a sensitive stage (primarily the third instar and the pupa) experienced a large increase (greater than 2 hrs.) in day length, from a relatively short to a long day (Tauber & Tauber 1976c). Similarly, diapause was terminated by a large increase in day length (Table 4-6B).

In the field, all individuals enter diapause immediately after emergence, because the sensitive stages are present just before and around the summer solstice, when day lengths are not increasing fast enough to avert diapause. Therefore, only one generation is produced each year. By testing diapausing adults from the field we showed that the insects' short-day requirement for diapause termination is satisfied during October (Tauber & Tauber 1975a). Subsequently, the rate of diapause termination is directly related to the duration of the prevailing day length, and diapause ends at the end of March.

Table 4-5. Preimaginal developmental rates of geographical populations in the *Chrysopa carnea* species-complex; mortality less than 10% except where indicated

	Days ($\bar{X}$ ± S. D.)				
	15.6°C	18.3°C	21.1°C	23.8°C	26.7°C
C. carnea, mohave strain, Strawberry Canyon, California	43.5 ± 1.9[a]	43.9 ± 0.1	32.3 ± 1.4	27.1 ± 0.2	21.3 ± 0.5
C. carnea, Goldstream Valley, Alaska	48.6 ± 2.9	35.5 ± 0.8	27.3 ± 0.9	20.3 ± 0.8	18.2 ± 1.5
C. carnea, Ithaca, New York	61.3 ± 1.9	43.3 ± 2.7	31.4 ± 0.9	26.3 ± 0.9	19.8 ± 0.4
C. carnea, El Sueco Junction Chihuahua, Mexico	62.5 ± 2.3[b]	41.9 ± 2.4	32.5 ± 0.4	24.1 ± 0.1	19.9 ± 0.9
C. downesi, St. Ignatius, Montana	78.6 ± 2.5	45.9 ± 1.5	35.7 ± 0.3	28.2 ± 0.6	22.3 ± 0.1
C. downesi, Ithaca, New York	76.9 ± 1.5	52.1 ± 0.9	36.8 ± 0.3	28.7 ± 0.2	22.0 ± 0.2

[a] Mortality = 72%.
[b] Mortality = 40%.

Table 4-6. Photoperiodic control of diapause in the *downesi*-type (short-day/long-day) life cycle[a] (T = 24 ± 1°C)

Diapause prevention/induction	
Photoperiodic conditions	Percentage diapause
L:D	
10:14	100
12:12	100
14:10	100
16:8	100
Increase in day length on day of spinning (prepupal stage)	
10:14→12:12	100
10:14→14:10	42
10:14→16:8	0
12:12→14:10	100
12:12→16:8	22
14:10→16:8	100
14:10→18:6	13

Diapause termination			
First photoperiodic condition (diapause-inducing L:D, from egg to 110-day-old adult)	Second photoperiodic condition (from day 111 to day 131)	Third photoperiodic condition (from day 132 onward)	Percentage oviposition
14:10	14:10	14:10	0
14:10	10:14	10:14	0
14:10	16:8	16:8	20
14:10	10:14	16:8	100
10:14	10:14	10:14	0
10:14	16:8	16:8	75

[a]Data from Ithaca, New York population (Tauber & Tauber 1976c).

The photoperiodic responses underlying the *carnea*- and the *downesi*- type life cycles are similar except for the short-day/long-day requirement for diapause avoidance and termination in the *downesi* type. This requirement acts to delay the expression of the quantitative response to day length. In the *carnea*-type life cycle, the quantitative response occurs in autumn and early winter; in the *downesi* type it occurs in winter and spring, after the short-day requirement is fulfilled. The short-day/long-day response underlying the *downesi* type life cycle is controlled by recessive alleles at each of two unlinked autosomal loci (Tauber et al. 1977). F_1 hybrids from crosses between *carnea*- and *downesi*-type individuals are all of the *carnea* type. The preoviposition period of hybrids is about 1-1.5 days longer than the pure *carnea* type, but it is 2-2.5 days shorter than the pure *downesi* type, as measured when diapause is averted (Tauber & Tauber 1977b).

Table 4-7. Photoperiodic control of the short-day/long-day/very long-day life cycle[a]

Diapause prevention/induction			
Photoperiodic conditions (L:D)			Percentage diapause
Egg-larval	Prepupa-pupal	Adult	
10:14	10:14	10:14	100
14:10	14:10	14:10	100
16:8	16:8	16:8	100
18:6	18:6	18:6	100
22:2	22:2	22:2	100
10:14	16:8	16:8	100
10:14	22:2	22:2	88
10:14	16:8	22:2	0
10:14	18:6	18:6	100
14:10	20:4	20:4	100

Diapause termination		
Photoperiodic conditions (L:D)		Percentage diapause termination[b]
Egg to 30-day-old adult	31-day-old adult onward	
16:8	23:1	100
16:8	24:0	100
18:6	23:1	100
18:6	24:0	100
20:4	23:1	0
20:4	24:0	87
22:2	23:1	0
22:2	24:0	57

[a]Data from Goldstream Valley, Alaska, population of *Chrysopa carnea* (T = 24 ± 1°C).
[b]Within 20 days of photoperiodic change.

The Alaska Type (Short-Day/Long-Day/Very Long-Day Reproduction). In addition to the *downesi* type short-day/long-day response, there is a second category, in which the insects respond to two increases in day length. In this case, diapause ensues under any constant day lengths tested, and also when the insects experience a single increase in day length (Table 4-7). However, a second increase, to a very long day length (Light: Dark 22:2), averts or terminates diapause (Table 4-7).

This type of photoperiodic requirement for diapause aversion and termination probably results in a univoltine life cycle, with reproduction delayed until near or after the summer solstice. We have not yet verified this life history variation by sampling natural populations in Alaska; however, we are analyzing the genetic mechanisms controlling the short-day/long-day/very long-day life cycle.

The Mohave-*Type (Prey-Mediated Reproduction).* The final type of disjunct variation in the *Chrysopa carnea* species-complex is a diapause mediated by the absence of food, which, under some conditions, reduces reproduction in the population. Lacewings with this type of variation possess the *carnea*-type diapause response to short day lengths, but they may also enter diapause under long days if prey become scarce (Tauber & Tauber 1973c). *Mohave*-type adults that experience long-day conditions (LD 16:8) and that receive aphids (*Myzus persicae*) and a protein and honey mixture as food reproduce without diapause; most of those reared and maintained under the same photoperiodic conditions, but receiving no aphids after emergence, enter reproductive diapause (Table 4-8).

The *mohave*-type life cycle is as follows: In late summer-early autumn, adults enter reproductive diapause in response to short day lengths. The photoperiodically controlled diapause continues until mid- to late winter, after which termination of diapause and initiation of reproduction depends on the availability of prey. Reproduction by subsequent generations even during the long days of mid- to late spring and summer depends on the availability of prey. If prey are scarce, adults enter an estival diapause that can be terminated by the presence of prey.

There is a disadvantage correlated with the ability to enter a prey-mediated diapause. Even in the presence of abundant prey, the preoviposition period of non-

Table 4-8. Dietary influence on diapause in the *mohave*-type life cycle[a]

Diapause prevention/induction		
Photoperiod (L:D)	Adult diet	Percentage diapause[b]
10:14	Wheast[c]	100
10:14	Wheast + aphids	100
12:12	Wheast	100
12:12	Wheast + aphids	100
14:10	Wheast	100
14:10	Wheast + aphids	100
16:8	Wheast	88
16:8	Wheast + aphids	0

Diapause termination		
Diapause-inducing L:D (egg to 21-day-old adult)	Diet after transfer to L:D 16:8 at day 22	Preoviposition period ($\bar{X} \pm$ S. D. days after transfer to L:D 16:8 on day 22[d])
12:12	Protein[e]	27.0 ± 0
12:12	Wheast + aphids	16.0 ± 0
10:14	Protein	28.0 ± 3.5
10:14	Wheast + aphids	16.0 ± 8.1
8:16	Protein	25.6 ± 4.0
8:16	Wheast + aphids	13.5 ± 3.5

[a] Data from Strawberry Canyon, California, population (Tauber & Tauber 1973c). (T = 24 ± 1°C.)
[b] No. ≥ 10 pairs.
[c] Wheast:sugar:water = 2:1:1.
[d] No. = 6-16.
[e] Protein hydrolysate of yeast (type A), sugar, and water.

diapause *mohave*-type females is significantly longer than that of *carnea*-type females under diapause-averting, long-day conditions (Table 4-9).

Hybridization between *carnea*-type (no estival diapause, no prey requirement) and *mohave*-type (prey-mediated estival diapause) individuals yields F_1 offspring that lack a prey requirement for reproduction and that never enter diapause under long-day conditions (Tauber & Tauber 1975b). Also, the preoviposition period of F_1 hybrids is similar to that of *carnea*-type females. Therefore, the *carnea*-type life cycle is dominant over the *mohave* type. Analysis of the inheritance of the prey-mediated estival diapause continues.

Geographical Variability in *Chrysopa*

We have studied the seasonal and life history traits of several populations in detail, and we have surveyed particular traits of many other populations throughout North America. In this section we describe local populations from four geographical areas and illustrate how their traits adapt them to meet the physical and biotic requirements of their local environments.

Northeastern United States—Ithaca, New York Populations

All of the populations of the *Chrysopa carnea* species-complex that we sampled from the northeastern United States fall into one of two types: the multivoltine *carnea*-type with long-day reproduction and short-day diapause, or the *downesi*-type univoltine life cycle with a short-day/long-day requirement for reproduction.

In the northeastern United States these types of life cycle occur in reproductively isolated taxa, with separate habitats and asynchronous seasonal cycles. The multivoltine *C. carnea* occurs in meadows and fields and on deciduous trees; its light green color blends in well with the vegetation in its habitat. Breeding takes place in early spring and throughout summer. At the end of summer, when day lengths decrease to the critical level, oviposition ceases and adults enter a reproductive diapause. Diapause induction is associated with adult movement from fields and meadows to forest edges where the dormant adults overwinter. This migration is accompanied by a color change from light green to yellowish or reddish brown, which provides crypsis against a background of senescent foliage and ground cover.

In contrast to *C. carnea*, the univoltine *C. downesi* is dark green during all phases of the adult stage. This species remains in the coniferous habitat throughout the year, and its adult coloration makes it difficult, at least for the human eye, to detect.

Breeding by *C. downesi* in the northeast occurs only in early spring. As a result, the seasonal cycles of the two sympatric *Chrysopa* species are asynchronous. That is, *C. downesi* is either immature or in reproductive diapause when *C. carnea* is reproductively active (Figure 4-2).

Laboratory-produced hybrids between the two species are intermediate in color, a hue that would be at a disadvantage in both the coniferous habitat and the field-meadow-deciduous tree situation. However, hybrids are not found in nature. For *C.*

Chrysopa downesi — Ithaca

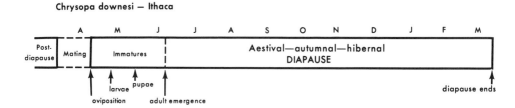

Chrysopa carnea — Ithaca

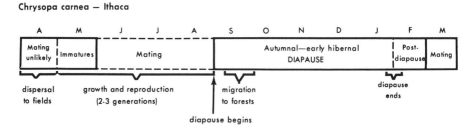

Figure 4-2. Seasonal cycles of *Chrysopa carnea* and *Chrysopa downesi* in the Ithaca, New York area. Note that the time of mating in the two species is seasonally asynchronous.

carnea and *C. downesi* in the northeast, the life history traits determining the seasonal timing and duration of reproduction serve not only as adaptations to the physical environment but also as a mechanism preventing interbreeding by two sympatric species.

We and others contend that the evolution of reproductive isolating mechanisms in sympatric populations, or in allopatric populations undergoing secondary contact, occurs relatively quickly (e.g. Bush 1974, Maynard Smith 1966, Tauber & Tauber 1981b). This is borne out by our data showing that the differential species-specific responses underlying asynchrony between *C. carnea* and *C. downesi* in the northeast are based on allelic differences at only two loci. We proposed that relatively simple genetic changes led to changes in the seasonal timing and expression of fundamental photoperiodic reactions; this, in turn, led to gross changes in the seasonal cycle and ultimately to reproductive isolation (Tauber & Tauber 1977a,b). It appears the *C. downesi's* univoltine life cycle, with its limited reproductive capacity, evolved relatively quickly from a multivoltine life cycle.

We conclude that two different types of selection pressure act on *C. carnea's* and *C. downesi's* life history traits. In *C. carnea* the timing of diapause induction, the thermal requirements for postdiapause development, and the thermal requirements for preimaginal growth and development are related to maximizing growth and reproduction within the physically favorable growing season. The length of the critical photoperiod generally exhibits a north-south gradient, allowing reproduction to continue late in the season at latitudes with long growing seasons. Postdiapause requirements are correlated with the severity and predictability of vernal conditions (Table 4-3), thus enabling early, but not premature, reproduction.

In contrast, *C. downesi's* life history traits are modified to subserve a very restricted breeding season which, in turn, subserves two possible functions. First, univoltinism with reproductive diapause during most of the summer prevents *C. downesi* from inter-breeding with *C. carnea*. Second, the restriction of immatures to spring probably avoids heavy larval mortality that would result from the low prey densities and high competi-tor densities that are associated with conifers in summer. Both the photoperiodic and the thermal reactions of *C. downesi* are modified to restrict reproduction. For example, the high t value for *C. downesi's* preimaginal growth and development serves to pro-long early larval development in the spring. This results in the sensitive stages occurring after day lengths are increasing rapidly enough to avoid diapause, and so estival dia-pause ensues, interbreeding with *C. carnea* is precluded, and larval occurrence is con-fined to spring (Tauber & Tauber 1976a).

Western United States–Strawberry Canyon, California Population

The population in Strawberry Canyon, Alameda County, California contains a mixture of characters that combines the *mohave-*, *carnea-*, and *downesi*-type life cycles. Unlike the two sympatric and reproductively isolated populations in the Northeast, this pop-ulation has not diverged into reproductively isolated units. Instead, the diverse charac-teristics are present in a single, interbreeding polymorphic population. The polymor-phism may result from selection for a "best-bet strategy" in an unpredictable environ-ment (see Istock 1978).

The Mediterranean climate of Strawberry Canyon presumably results in drastic between-year differences in prey availability for the *Chrysopa* larvae. Depending on the temperature conditions and the occurrence, amount, and timing of the winter and early spring rainfall, prey occurrence can vary from late winter to late spring. Less prey is probably available in summer than in spring, and the amount of prey varies between years. The population in Strawberry Canyon has evolved a mechanism that closely "tracks" the prey. Autumnal photoperiods induce and maintain diapause, but diapause termination and the initiation of reproduction in late winter-early spring is timed by the occurrence of prey. Likewise, reproduction during summer is directly dependent on the availability of prey. Field collections over many years and exam-ination of museum specimens indicate that the percentage of reproductively active adults (as opposed to estivating adults) in July and August ranges from 15 to 73% depending on the year (Tauber & Tauber 1973c).

In addition to "tracking" the availability of prey, the population of Strawberry Canyon responds to an increase in day length. In the laboratory, an increase in day length during a late larval stage results in 100% oviposition, even in the absence of prey (Table 4-9). This response accounts for reproduction by 100% of the population each spring. However, the preoviposition period in females that received both an increase in day length and a supply of aphids is significantly shorter than in those that received only an increase in day length (Table 4-9).

Adults of the Strawberry Canyon population are a slightly lighter green than adults of *C. carnea* in the northeastern United States. Strawberry Canyon larvae and adults also have distinct facial and body markings (Tauber 1974). This population

Table 4-9. Time to initiate oviposition by the *C. carnea, mohave* population from Strawberry Canyon, California[a]

L:D (egg to prepupa)	L:D (prepupa onward)	Aphid supplement to adult diet	Percentage diapause[b]	Preoviposition period ($\overline{X} \pm$ S.D. days)[b]
10:14	16:8	+	0	8.4 ± 1.9[c]
10:14	16:8	0	0	12.6 ± 3.3
16:8	16:8	+	30	13.6 ± 3.0
16:8	16:8	0	98	ca. 90 days (but occasional)
10:14	10:14	+	100	ca. 60 days (but occasional)
10:14	10:14	0	100	Over 200 days (but rare)

[a]From Tauber and Tauber (1973c, 1981a).
[b]No. in each condition = 14-58.
[c]Cf. nondiapause preoviposition period of *C. carnea*, Ithaca, New York, population = 4.3 ± 0.6.

undergoes a color change from light green to yellowish or reddish brown, similar to *C. carnea* in the northeast. Adults spend most of the year on the deciduous evergreen trees *Umbellularia californica* and *Quercus agrifolia*. However, in spring they also occur on shrubs.

Currently, we are investigating isofemale lines to determine the genetic basis for the polymorphic response to prey. We are also testing for homology between the Strawberry Canyon population's and *C. downesi's* short-day/long-day requirement for reproduction.

Northwestern United States—St. Ignatius, Montana Population

We analyzed a population of dark green adults from St. Ignatius, Montana. This population, which was derived from adults collected from an overwintering site near evergreen trees, bred true for dark green. A light green form of *C. carnea* may also occur in the area; based on collection records we presume that it does.

The life history traits of this population resemble those of *C. downesi* from the Northeast (Tauber & Tauber 1981a). The population shows a short-day/long-day requirement for reproduction, diapausing adults remain dark green, and the thermal requirements for development are relatively high (Tables 4-4, 4-5, and 4-10). We propose, therefore, that like the eastern population this population is univoltine. Its estival-autumnal-hibernal diapause is induced primarily by, and persists in response to, photoperiod until late winter. Like the northeastern population, the high thermal requirements for preimaginal development of the St. Ignatius population probably subserve the maintenance of univoltinism.

However, this population differs from eastern *C. downesi* in that prey presence slightly reduces diapause incidence and greatly shortens the preoviposition period.

Table 4-10. Diapause and life history characteristics of *C. downesi* from St. Ignatius, Montana, and Ithaca, New York (T = 24 ± 1°C)[a]

L:D (egg to prepupa)	L:D (prepupa onward)	Aphid supplement to adult diet	Percentage diapause[b]	Preoviposition period ($\overline{X}$ ± S.D. days)[b]
St. Ignatius, Montana				
10:14	16:8	+	0	13.2 ± 2.4
10:14	16:8	0	0	20.7 ± 5.7
16:8	16:8	+	95	27
16:8	16:8	0	100	–
Ithaca, New York				
10:14	16:8	+	0	9.2 ± 2.3
10:14	16:8	0	0	10.2 ± 2.4
16:8	16:8	+	100	–
16:8	16:8	0	100	–

[a] From Tauber and Tauber (1976c, 1981a).
[b] No. in each condition = 8-24.

Therefore, the timing of the initiation of oviposition in the spring depends on prey availability at that time (Tauber & Tauber 1981a). This dual mechanism for controlling reproductive activity indicates that the St. Ignatius population combines the *downesi*-type life cycle with aspects of the *mohave* type.

Northern United States–Goldstream Valley, Alaska Population

The population we will discuss last is one from 9 miles northwest of Fairbanks, in Goldstream Valley, Alaska. Several characteristics distinguish it from the populations described above: It has a short-day/long-day/very long-day requirement but no prey requirement for reproduction. It needs relatively few heat-degree days (K) for initiating postdiapause oviposition, but the threshold (t) is relatively high. Its thermal requirements for preimaginal growth and development are much lower than those for the other populations studied (Table 4-4). Also, the Goldstream Valley population is brown most of the year, and it does not assume its full green coloration until it is fully reproductive.

We conclude that this population is univoltine. Its photoperiodic requirements for reproduction insure that oviposition will not begin until day lengths are very long. This characteristic, in combination with the very high threshold (t) for postdiapause development, serves to delay ovarian maturation until late in the season. These adaptations interact to preclude premature reproduction in an area that can experience severe cold periods in late spring. However, once ovarian development begins, it proceeds very rapidly; i.e., the K values are very low. This also holds for preimaginal development. This population's reproduction and development therefore are adapted to a very short growing season.

Perhaps because of the fast growth rate, adults of the Goldstream Valley population are slightly smaller than those from southern populations. This suggests a "trade-off" between reproductive potential and growth rates.

Conclusion

A variety of diverse, often opposing selective forces act on the life history traits of the *Chrysopa carnea* species-complex (Table 4-11). Populations are characterized by co-ordinated sets of life history traits that adapt them to the physical and biotic factors in their particular locality and habitat. Selection for one character (e.g., photoperiodically controlled univoltinism) can strongly affect numerous other life history traits (e.g., growth rate, fecundity, and oviposition schedule).

In some instances, as in the Alaskan population, adaptation to a very harsh physical environment dominates the shaping of the life cycle and the underlying life history traits. Under the milder conditions of Strawberry Canyon, California, where there is considerable variability in the availability of food, reproduction is keyed to the seasonal occurrence of prey. In this case, the life history traits subserve a polymorphic response to variable prey levels and constitute a "best-bet" strategy in an unpredictable environment. Similarly, seasonal occurrence of prey also influences the timing of vernal reproduction in the univoltine population from St. Ignatius, Montana.

In the sympatric populations of *C. carnea* and *C. downesi*, diverse selection pressures appear to influence the evolution of life history traits. In one of the species, *C. downesi*, the life history traits are adapted to prevent interbreeding with the more common and dominant *C. carnea*. In the other species, *C. carnea*, the life history traits are not influenced by the need for reproductive isolation and are primarily adapted to the local climatic conditions.

Several genetic mechanisms underlie the geographical variability in life history traits of the *Chrysopa carnea* species-complex. In one situation, the photoperiodic responses that severely limit reproduction (*C. downesi's* univoltinism) result from a simple Mendelian system, involving recessive alleles at two autosomal loci. More complicated genetic systems, involving several genes with dominace and recessiveness, regulate other traits, such as the prey-mediated estival diapause in the Strawberry Canyon, California population. Still other traits, such as the geographically variable critical photoperiod for diapause induction in *C. carnea*, come under the control of multiple genes with additive affects.

These genetic systems underlie both continuous and disjunct variation is seasonal and life history traits. Expression of genetic variation in the *Chrysopa carnea* species-complex runs the full range from continuous variation in quantitative characters (e.g., the critical photoperiod for *C. carnea* diapause induction) to the polymorphic expression of alternate life cycles in a single population, and finally to the expression of fully asynchronous life cycles in two sympatric, potentially interbreeding but seasonally isolated species.

Acknowledgments. We gratefully acknowledge Dr. J. R. Nechols, presently at the University of Guam, and J. J. Obrycki, Cornell University, for their sustained help. The following individuals contributed generously to our study: Drs. L. A. Falcon,

Table 4-11. Summary of life history and seasonal cycle variation in *C. carnea* species-complex

I. Quantitative variation
    A. Critical photoperiod for diapause induction
    B. Depth of diapause
    C. Thermal requirements for initiation of postdiapause oviposition
    D. Thermal requirements for preimaginal development
    E. Influence of prey availability on non-diapause preoviposition period

II. Disjunct Variation
    A. Long-day reproduction (*carnea* type)
       1. Multivoltine
       2. No estivation
       3. Diapause averted by long day lengths, induced by day lengths below critical photoperiod
       4. Diapause maintained by day length until midwinter; no specific terminating stimulus
       5. No dietary requirement for reproduction

    B. Short-day/long-day reproduction (*downesi* type)
       1. Univoltine; reproduction in spring
       2. Photoperiodically controlled estival-autumnal-hibernal diapause
       3. Diapause averted (in laboratory) by large increase in day length
       4. Diapause termination timed by photoperiod
       5. No dietary requirement for reproduction

    C. Short-day/long-day/very long-day reproduction (Alaska type)
       1. Univoltine
       2. Photoperiodically controlled late estival-autumnal-hibernal diapause
       3. Diapause averted (in laboratory) by two large increases in day length to a very long day
       4. Diapause termination probably timed by photoperiod
       5. No dietary requirement for reproduction

    D. Prey-mediated reproduction (*mohave* type)
       1. Voltinism variable
       2. Prey-mediated estival diapause
       3. Estival diapause averted by abundance of prey
       4. Termination of overwintering diapause timed by availability of prey
       5. Prey-mediated diapause only expressed under diapause-averting photoperiods

J. G. Franclemont, G. L. Jensen, C. D. Johnson, M. K. Kennedy, K. W. Philip, J. A. Powell, G. D. Propp, (Mrs.) O. Smith, G. Tamaki, and W. H. Whitcomb. We thank Professors J. Antonovics, Duke University, and H. Dingle, University of Iowa, for their comments on the manuscript. Our work was supported by National Science Foundation grants DEB-7725486 and DEB-8020988.

References

Alrouechdi, K., Canard, M.: Mise en évidence d'un biotype sans diapause photopériodique dans une population méditerraneenne de *Chrysoperla carnea* (Stephens) (Insectes, Neuroptera). C.R. Acad. Sci. Paris Ser. D. 289, 553-555 (1979).

Ando, Y.: Geographic variation in the incidence of non-diapause eggs of the false melon beetle, *Atrachya menetriesi* Falderman (Coleoptera: Chrysomelidae). Appl. Entomol. Zool. 14, 193-202 (1979).

Blackman, R. L.: Variation in the photoperiodic response within natural populations of *Myzus persicae* (Sulz.). Bull. Entomol. Res. 60, 533-546 (1971).

Blackman, R. L.: Life-cycle variation of *Myzus persicae* (Sulz.) (Hom., Aphididae) in different parts of the world, in relation to genotype and environment. Bull. Entomol. Res. 63, 595-607 (1974).

Bradshaw, W. E.: Homeostasis and polymorphism in vernal development of *Chaoborus americanus*. Ecology 54, 1247-1259 (1973).

Bush, G. L.: The mechanism of sympatric host race formation in the true fruit flies. In: Genetic Mechanisms of Speciation in Insects, White, M. J. D., (ed.). Boston: D. Reidel, 1974, pp. 3-23.

Danilevskii, A. S.: Photoperiodism and Seasonal Development of Insects. (English translation) London: Oliver & Boyd, 1965.

Denno, R. F., Grissell, E. E.: The adaptiveness of wing-dimorphism in the salt marsh-inhabiting planthopper, *Prokelisia marginata* (Homoptera: Delphacidae). Ecology 60, 221-236 (1979).

Dingle, H.: The experimental analysis of migration and life-history strategies in insects. In: Experimental Analysis of Insect Behaviour. Barton Brown, L. (ed.). New York: Springer-Verlag, 1974, pp. 329-342.

Dingle, H.: Migration and diapause in tropical, temperate, and island milkweed bugs. In: Evolution of Insect Migration and Diapause. Dingle, H. (ed.). New York: Springer-Verlag, 1978, pp. 254-276.

Dingle, H., Brown, C. K., Hegmann, J. P.: The nature of genetic variance influencing photoperiodic diapause in a migrant insect, *Oncopeltus fasciatus*. Am. Nat. 111, 1047-1059 (1977).

Goldschmidt, R.: *Lymantria*. Biblio. Genet. 11, 1-186 (1934).

Hagen, K. S., Tassan, R. L.: The influence of protein hydrolysates of yeasts and chemically defined diets upon the fecundity of *Chrysopa carnea* Stephens (Neuroptera). Acta Soc. Zool. Bohemoslov. 30, 219-227 (1966).

Istock, C. A.: Fitness variation in a natural population. In: Evolution of Insect Migration and Diapause. Dingle, H. (ed.). New York: Springer-Verlag, 1978, pp. 171-190.

Istock, C. A., Zisfein, J., Vavra, K. J.: Ecology and evolution of the pitcher-plant mosquito. 2. The substructure of fitness. Evolution 30, 535-547 (1976).

Kidokoro, T., Masaki, S.: Photoperiodic response in relation to variable voltinism in the ground cricket, *Pteronemobius fascipes* Walker (Orthoptera: Gryllidae). Jap. J. Ecol. 28, 291-298 (1978).

Krysan, J. L., Branson, T. F., Castro, G. D.: Diapause in *Diabrotica virgifera* (Coleoptera: Chrysomelidae): A comparison of eggs from temperate and subtropical climates. Entomol. Exptl. Appl. 22, 81-89 (1977).

Lees, A. D.: The physiology of Diapause in Arthropods. London: Cambridge University Press, 1955.

Lumme, J., Oikarinen, A.: The genetic basis of the geographically variable photoperiodic diapause in *Drosophila littoralis*. Hereditas 86, 129-142. (1977).

MacLeod, E. G.: Experimental induction and elimination of adult diapause and autumnal coloration in *Chrysopa carnea* (Neuroptra). J. Insect Physiol. 13, 1343-1349 (1967).

Masaki, S.: Geographic variation of diapause in insects. Bull. Fac. Agr. Hirosaki Univ. 7, 66-98 (1961).

Masaki, S.: Seasonal and latitudinal adaptations in the life cycle of crickets. In: Evolution of Insect Migration and Diapause. Dingle, H. (ed.). New York: Springer-Verlag, 1978, pp. 72-100.

Masaki, S.: Climatic adaptation and species status in the lawn ground cricket. I. Photoperiodic response. Kontyû 47, 48-65 (1979).

Maynard Smith, J.: Sympatric speciation. Am.Nat. 100, 637-650 (1966).

Obrycki, J. J., Tauber, M. J.: Phenology of three coccinellid species: thermal requirements for development. Ann. Entomol. Soc. Am. 74, 31-36 (1981).

Sheldon, J. K., Macleod, E. G.: Studies on the biology of the Chrysopidae II. The feeding behavior of the adult of *Chrysopa carnea* (Neuroptera). Psyche Cambr. 78, 107-121 (1971).

Sims, S. R.: Diapause dynamics and host plant suitability of *Papilio zelicaon* (Lepidoptera: Papilionidae). Am. Midl. Nat. 103, 375-384 (1980).

Southwood, T. R. E.: Habitat, the templet for ecological strategies? J. Anim. Ecol. 46, 337-365 (1977).

Stearns, S. C.: Life-history tactics: A review of the ideas. Q. Rev. Biol. 51, 3-47 (1976).

Tauber, C. A.: Systematics of North American chrysopid larvae: *Chrysopa carnea* group (Neuroptera). Can. Entomol. 106, 1133-1153 (1974).

Tauber, M. J., Tauber, C. A.: Geographic variation in critical photoperiod and in diapause intensity of *Chrysopa carnea* (Neuroptera). J. Insect Physiol. 18, 25-29 (1972).

Tauber, M. J., Tauber, C. A.: Seasonal regulation of dormancy in *Chrysopa carnea* (Neuroptera). J. Insect Physiol. 19, 1455-1463 (1973 a).

Tauber, M. J., Tauber, C. A.: Quantitative response to daylength during diapause in insects. Nature (London) 244, 296-297 (1973 b).

Tauber, M. J., Tauber, C. A.: Nutritional and photoperiodic control of the seasonal reproductive cycle in *Chrysopa mohave* (Neuroptera). J. Insect Physiol. 19, 729-736 (1973 c).

Tauber, M. J., Tauber, C. A.: Natural daylengths regulate insect seasonality by two mechanisms. Nature (London) 258, 711-712 (1975 a).

Tauber, M. J., Tauber, C. A.: Criteria for selecting *Chrysopa carnea* biotypes for biological control: Adult dietary requirements. Can. Entomol. 107, 589-595 (1975 b).

Tauber, M. J., Tauber, C. A.: Environmental control of univoltinism and its evolution in an insect species. Can. J. Zool. 54, 260-266 (1976 a).

Tauber, M. J., Tauber, C. A.: Insect seasonality: Diapause maintenance, termination and postdiapause development. Ann. Rev. Entomol. 21, 81-107 (1976 b).

Tauber, M. J., Tauber, C. A.: Developmental requirements of the univoltine *Chrysopa downesi*: Photoperiodic stimuli and sensitive stages. J. Insect Physiol. 22, 331-335 (1976 c).

Tauber, C. A., Tauber, M. J.: Sympatric speciation based on allelic changes at three loci: Evidence from natural populations in two habitats. Science 197, 1298-1299 (1977 a).

Tauber, C. A., Tauber, M. J.: A genetic model for sympatric speciation through habitat diversification and seasonal isolation. Nature (London) 268, 702-705 (1977 b).

Tauber, M. J., Tauber, C. A.: Evolution of phenological strategies in insects: A comparative approach with eco-physiological and genetic considerations. In: Evolution of Insect Migration and Diapause. Dingle, H. (ed.). New York: Springer—Verlag, 1978, pp. 53-71.

Tauber, M. J., Tauber, C. A.: Seasonal responses and their geographic variation in *Chrysopa downesi*: Ecological and evolutionary considerations. Can. J. Zool. 59, 370-376 (1981 a).

Tauber, C. A., Tauber, M. J.: Insect seasonal cycles: genetics and evolution. Ann. Rev. Ecol. Syst. 12, 281-308 (1981 b).

Tauber, M. J., Tauber, C. A., Denys, C. J.: Adult diapause in *Chrysopa carnea:* Photoperiodic control of duration and colour. J. Insect Physiol. 16, 949-955 (1970).

Tauber, C. A., Tauber, M. J., Nechols, J. R.: Two genes control seasonal isolation in sibling species. Science 197, 592-593 (1977).

Vepsäläinen, K.: Determination of wing length and diapause in water-striders (*Gerris* Fabr., Heteroptera). Hereditas 77, 163-177 (1974).

Vepsäläinen, K.: Wing dimorphism and diapause in *Gerris*: Determination and adaptive significance. In: Evolution of Insect Migration and Diapause. Dingle, H. (ed.). New York: Springer-Verlag, 1978, pp. 218-253.

Waldbauer, G. P.: Phenological adaptation and the polymodal emergence patterns of insects. In: Evolution of Insect Migration and Diapause. Dingle, H. (ed.). New York: Springer-Verlag, 1978, pp. 127-144.

PART THREE

Modes of Reproduction

One of the important aspects of an organism's life history is the way it reproduces. Demographic schedules and the effectiveness of forces acting on gene frequencies are both subject to the influence of reproductive strategies. The variety of these strategies and their implications for life history evolution are addressed in the four chapters that constitute this section.

Alan Templeton explores the genetic basis and genetic implications of parthenogenesis using a species from the Hawaiian *Drosophila* complex, *D. mercatorum*. Parthenogenesis imposes profound changes in life history traits, and Templeton argues that these changes, rather than parthenogenesis itself, are instrumental in differential success among isofemale clones. R. Jack Schultz uses comparisons among sexual, semisexual, and asexual biotypes of the viviparous fish genus *Poeciliopsis* to examine the relative success of various forms (clones and hemiclones) with respect to growth rates and fecundity. Both papers demonstrate that isogenicity can be a highly successful strategy and both cast doubt on the necessity for maintaining sexual reproduction to avoid genetic "dead ends."

Charles King uses rotifer clones to consider a major problem in demographic theory, the evolution of life span. In these asexually reproducing organisms, older individuals contribute little to fitness by the further production of offspring but they do influence the production of grandchildren. King argues that selection should act to adjust life span to twice the generation time, based on contributions to offspring and grand-offspring. The final contribution to this section, by Thomas Meagher and Janis Antonovics, evaluates the relative importance of life history traits in each sex of the dioecious plant *Chamaelirium luteum*. Noting that sex is the most common genetic polymorphism, these authors consider the consequences for demography in the contrast between male and female life histories. The two papers illustrate interesting alternative ways to use species of various reproductive modes in addressing issues in population biology.

Chapter 5

The Prophecies of Parthenogenesis

ALAN R. TEMPLETON*

Introduction

Parthenogenesis is the development of an unfertilized egg into a new individual. This phenomenon has long fascinated both scientists and laymen alike. Undoubtedly, the most celebrated case of a virgin birth is that given in Matthew 1, 19-25: the account of the birth of Jesus Christ. Whether or not one accepts this account literally, one can still legitimately ask the question: Why is this account of a virgin birth contained in Matthew? Most people would respond by answering there was a prophecy that the Messiah would be the result of a virgin birth, and that this account was inserted into the Gospel to show that this prophecy was met. If the person asked is a bit more knowledgeable about the Bible, he or she may quote Isaiah 7, 14, which is also quoted in Matthew 23 and reads in the King James version: "Behold, a virgin shall conceive, and bear a son, and shall call his name Immanuel." Thus, the prophecy seems to be rather clear-cut. However, if the same passage in Isaiah is read in the New English Bible, it begins: "A young woman is with child." The word in question here is the Hebrew word *almah* which scholars of the Hebrew language agree simply means a young woman of marriageable age, whether married or not (Argyle 1963, p. 28). However, the quote of Isaiah 7, 14 found in Matthew gives the Greek word *parthenos* meaning "virgin" as the translation of *almah*. Consequently, this most famous instance of parthenogenesis does not owe its existence to any "prophecy of a virgin birth," for such a prophecy never existed. Instead, it owes its existence to the far more mundane reason of a mistranslation from Hebrew to Greek (Argyle 1963, p. 28).

In less celebrated instances of parthenogenesis, other "prophecies of parthenogenesis" have been invoked to explain this phenomenon. For example, the "prophecy of environmental uncertainty vs. environmental certainty," the "prophecy of weedy habitats," and the twin genetic-ecological "Prophecies of the cost of meiosis and the cost of sex" have been frequently invoked, just to mention a few. These prophecies are grand sounding and certainly impart a profound reason for the evolution of parthenogenesis; but are these prophecies real, or, as in the case of Jesus, are there more mun-

*Department of Biology, Washington University, St. Louis, Missouri 63130 U.S.A.

dane, less flashy explanations for the origin of parthenogenesis? I will argue that the latter reasons do exist but can be uncovered only by examining the genetic basis and genetic implications of parthenogenetic reproduction and by constantly remembering that, because evolution is the genetic transformation of a population through time, evolutionary outcomes are always severely constrained by the nature of the genetic systems involved and by the developmental mechanisms through which this genetic system exerts an influence on the phenotype.

In particular, I will support this argument with genetic studies on parthenogenesis in the genus *Drosophila*. I also intend to show how these *Drosophila* studies can provide much insight into the evolutionary origin of parthenogenetic species of insects in general, although my discussion will be limited to cases of thelytoky (female-producing parthenogenesis) of nonhybrid origin.

Tychoparthenogenesis and the Abandonment of Sex

The pioneering work of Stalker (1951, 1952, 1954) first established that parthenogenesis existed in the genus *Drosophila* and, moreover, that it was a widespread phenomenon. Since Stalker's initial survey, many other *Drosophila* have been shown to display tychoparthenogenesis, that is, accidental or rare parthenogenesis in an otherwise sexually reproducing species (Carson 1961, 1962, 1967a, Futch 1973, Templeton 1979a). In addition, this capacity for tychoparthenogenesis is more pronounced in natural populations than in inbred laboratory populations (Carson 1961, Templeton 1979a, Templeton et al. 1976a), thereby implying parthenogenesis is a true evolutionary potential in the genus. Only one species, *Drosophila mangabeirai* (Carson et al. 1957), has utilized this potential fully and become totally parthenogenetic.

All parthenogenetic *Drosophila* are automictic; that is, they retain normal meiosis, and most are diploid, although polyploidy frequently arises. Because most are diploid and have normal meiosis, the question immediately arises as to how diploidy is restored in the absence of fertilization. This is accomplished by one of three mechanisms in *Drosophila*. The first is central fusion (Figure 5-1) in which two haploid pronuclei that segregated at meiosis I fuse to restore diploidy. As can be seen from this figure, central fusion also restores the maternal genotypic composition for all loci that have not recombined with their centromere and even restores the maternal state half the time when there is recombination. The decay into homozygosity is slow under central fusion in organisms such as *Drosophila* with chromosomes of small recombinational length. Moreover, with inversions or other crossover suppressors yielding absolute linkage to the centromere, permanent heterozygosity can be maintained (Carson 1967b).

The second mechanism for restoring diploidy is terminal fusion (Figure 5-2) of pronuclei that divided at meiosis II which causes a rapid decay into homozygosity in the absence of selection. The final mechanism is called gamete or pronuclear duplication (Figure 5-3) in which a haploid pronucleus mitotically forms cleavage nuclei which then fuse. This mechanism enforces total homozygosity in a single generation.

At this point, some might already object to my earlier statement that parthenogenetic *Drosophila* can serve as a useful general model for the nonhybrid evolution of thelytoky in the insects. Automixis has often been dismissed from consideration of the evolution of parthenogenesis for both theoretical and factual reasons. For example,

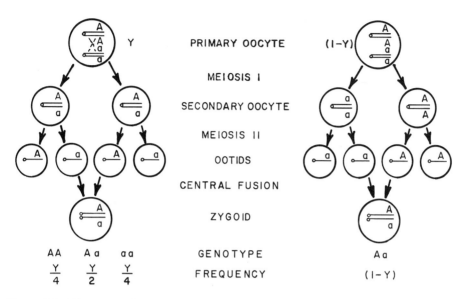

Figure 5-1. Diagrammatic representation of the meiotic events of central fusion. A single locus with two alleles (A and a) is considered, where y is the probability of a recombination between the locus and its centromere. From Asher (1970).

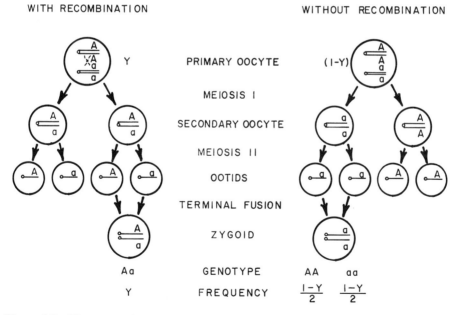

Figure 5-2. Diagrammatic representation of the meiotic events of terminal fusion. A single locus with two alleles (A and a) is considered, where y is the probability of a recombination between the locus and its centromere. From Asher (1970).

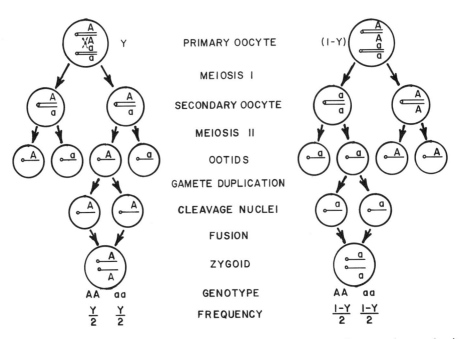

Figure 5-3. Diagrammatic representation of the meiotic events of gamete (pronuclear) duplication. A single locus with two alleles (A and a) is considered, where y is the probability of a recombination between the local and its centromere.

if one believes in the "prophecy of the cost of meiosis," it makes little sense to abandon sex but retain meiosis. In Maynard Smith's (1978) book on the evolution of sex, he gives other prophecies as to why automixis should be "an unpromising starting point for the origin of a parthenogenetic variety." Moreover, Maynard Smith notes that weevils are the only group of insects in which thelytoky is at all common and that these weevils are apomictic; that is, they reproduce parthenogenetically by suppressing meiosis. However, Maynard Smith failed to note that Seiler (1947) long ago showed that these weevils most likely evolved from automictic ancestors. More recently, Lokki and Saura (1980) have summarized evidence that diploid automixis is the major route in the evolution of apomictic polyploidy in insects. Therefore, the evolutionary causes for abandonment of sex in most apomictic insect populations can only be understood in terms of automixis, with apomixis being a derived, secondary character not directly involved in the reproductive transition (but perhaps very important in maintaining the long-term survival of the thelytokus population). In addition, naturally occurring automictic populations have evolved independently many times in the insects (Suomalainen et al. 1976), and therefore the phenomenon does seem to deserve some sort of evolutionary explanation. Finally, and most importantly, almost all cases of tychoparthenogenesis (i.e., accidental or rare parthenogenesis occurring in an otherwise sexually reproducing species) in insects have proved to be automictic. Maynard Smith (1978, p. 43) recognizes·this fact, but states (p. 63), "Cases in which an occasional egg develops without fertilization are not relevant, particularly if such development is auto-mictic." However, Maynard Smith fails to explain how thelytoky of nonhybrid origin can arise from a bisexual ancestor without tychoparthenogenesis. Prophecies to the contrary, the necessary prerequisite for the evolution of a thelytokous population is the existence of a tychoparthenogenetic capacity in the bisexual ancestors, which, at least in the insects, is most likely to be automictic. Any model for the evolution of insect parthenogenesis must be consistent with these constraints.

Any explanation for the evolution of thelytoky therefore must begin with auto-mictic tychoparthenogenesis in an otherwise sexual species. As soon as this is accepted, however, the experimental facts concerning tychoparthenogenesis immediately come into conflict with yet another prophecy of parthenogenesis, the "prophecy of the cost of sex." Basically, Maynard Smith (1978) and others have argued that thelytokous par-thenogenetic forms have an inherent twofold increase in rate of population growth over their sexual ancestors because all their offspring are females rather than just half being females as in most sexual species. However, observed rates of tychopartheno-genesis in natural populations of *Drosophila* vary from one egg in 100,000 to one egg in a million successfully developing into an adult (Templeton 1979a, Templeton et al. 1976a). Although this rate increases very rapidly once parthenogenetic reproduction has been initiated, it still remains well below 10% of the eggs successfully developing in the initial generations, and this seems to be a general property of tychoparthenogen-esis in other organisms as well (Lamb and Willey 1979). In addition, *D. mangabeirai* is a parthenogenetic species that has obviously been selected for parthenogenetic devel-opment for quite some time, yet only 60% of the eggs hatch, and even under optimal laboratory conditions only 80% of these survive to adulthood (Carson 1962). Hence, this long-established and successful parthenogenetic species (for it apparently drove its sexual ancestors to extinction) produces about the same number of female offspring as sexual *Drosophila* for a given number of eggs laid. Therefore, it is hard to see the relevance of the hypothesized doubled rate of growth to the initial reproductive tran-sition. Indeed, the experimental work and observed tychoparthenogenetic rates strongly suggest the real question should be how a parthenogentic population can evolve in spite of a several-fold disadvantage in the production of viable eggs?

One solution to the above problem, suggested by Stalker (1956a), is that virgin females occasionally may find themselves in a situation where locating a mate is diffi-cult or impossible: for example, a founder event involving only virgin females; or a part of a population being in an ecologically marginal habitat in which density is so low that not finding a mate is probable; or a species subdivided into small, semiisolated populations in which sampling variation in regard to the sex ratio has created local shortages or even total absences of males. Genetic situations can also arise that force females into the abandonment of sexual reproduction. For example, screens for par-thenogenesis in natural populations of *Drosophila mercatorum* have revealed the existence in the sexual population of an X-chromosomal element and an autosomal element which when jointly homozygous cause an almost complete destruction of female sexual receptivity. Obviously, such a gene complex normally would be selected against under sexual reproduction, but there might be other pleiotropic effects asso-ciated with heterozygosity or other genetic backgrounds that maintain such alleles in a polymorphic state, as they apparently were in this sexual population of *D. merca-torum*. Hence, some populations might regularly segregate out females with no sexual receptivity and which therefore could utilize a parthenogenetic capacity if such a capacity were present. More generally, Paterson (1978) has argued that many sexual species have a mate recognition system that includes visual, olfactory, tactile, etc., cues. Any mutation or gene complex causing significant changes in the cues or the perception of these cues easily could foreclose the option of sexual reproduction, and indeed Paterson has argued that such mate recognition systems would normally be subject to intense stabilizing selection. When coupled with tychoparthenogenesis, however, mutations or gene complexes disrupting the mate recognition system could also invoke selection for parthenogenetic reproduction.

Finally, some components of the mate recognition system in many organisms, including *Drosophila* (Pruzan et al. 1979, Sene 1977) are in part results of imprinting or cultural inheritance. If a female imprints by accident on the wrong mate recognition system, or perhaps on other females rather than males, once again a situation has arisen in which parthenogenesis is the only reproductive option. In summary, I believe the initial impetus for parthenogenetic reproduction is to be found in those geographical, ecological, and genetic circumstances that force the females into utilizing a tychoparthenogenetic capacity as their only reproductive option. More often than not, parthenogenesis is thrust upon the population, rather than being immediately optimal by some decision-theoretic criterion.

One attribute about the geographical, ecological, and genetic situations described above is that they tend to be rare but recurring. Recurrence is important in light of the very low initial efficiencies of tychoparthenogenetic reproduction, because the most common outcome of these recurring events is simply that the virgin females leave no offspring at all. However, an underlying tychoparthenogenetic capacity occasionally will be realized in the long run. Even at this initial stage, natural selection can occur that can leave a genetic imprint on the newly arisen parthenogenetic population throughout its entire evolutionary history. For example, it is evident from parthenogenetic screens of natural populations of *D. mercatorum* and *D. hydei* (Templeton 1979a, Templeton et al. 1976a) that virgin females who lay many eggs have a higher probability of leaving parthenogenetic progeny than virgins who lay few eggs. This is simply because each unfertilized egg has a very low probability of developing during the initial transitional generation. Hence, one tendency that accompanies the establishment of thelytoky is an increase in egg-laying capacity and its attendant pleiotropic trade-offs. For example, Figure 5-4 shows the egg-laying history of virgin daughters of wild-caught females under laboratory conditions. Moreover, the experiment was terminated when one-half of the original cohort died, so the length of the experiment gives the median longevity after eclosion. Figure 5-5 gives the same life history data for a parthenogenetic strain derived from these virgins. Note that the parthenogenetic females begin laying eggs after 2 days rather than 4, that they reach a peak egg production almost immediately that is higher than the peak for the sexual females, and that the median longevity is only 17 days as opposed to 54 days. Needless to say, these are rather dramatic life history differences, and any ecologist is likely to agree that the parthenogenetic strain displays a more "r-selected" phenotype. However, this bias toward such a life history arises quite simply from the selective forces associated with the initial low efficiency of tychoparthenogenetic reproduction. Of course, the survival of the parthenogenetic population may well depend on "r" or "weedy"

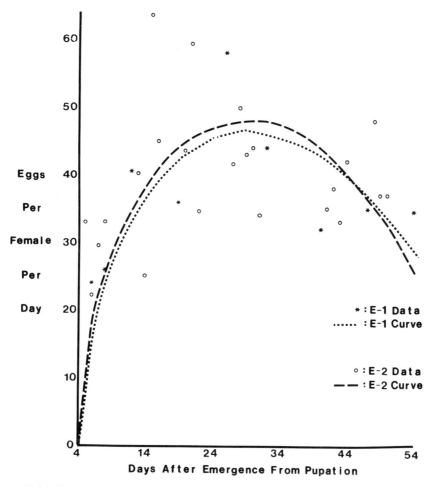

Figure 5-4. The number of eggs laid daily by virgin females from the sexual Kamuela stock of *Drosophila mercatorum*. Two different experimental designs were used (E-1 and E-2) and the curves for each were estimated using least squares. The experiment was terminated when one-half of the initial cohort had died. See Templeton et al. (1976a) for details.

environments being available so that some benefit accrues to the r-selected phenotypes that are forced upon the parthenogens. Consequently, parthenogenesis is not necessarily optimal in weedy environments, but weedy environments may be optimal for parthenogens.

In *D. mercatorum*, there is direct evidence for selection favoring genes associated with increased egg-laying capacity. The gene complex in question is known as abnormal abdomen, for in some combinations it yields adult flies with abdomens lacking bristles, pigmentation, and distinct tergites and sternites. In addition, the expression of abnormal abdomen is temperature dependent, disappearing totally when the larvae are raised at $17-18°C$, and it behaves as a lethal to semilethal complex depending upon culture

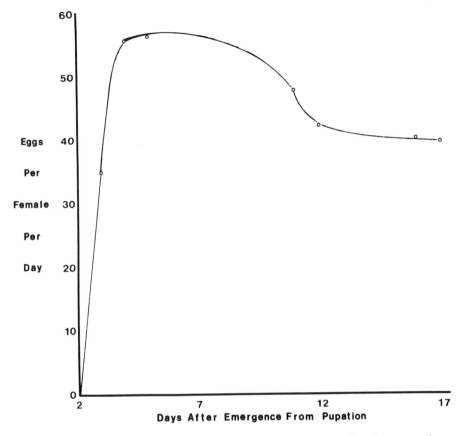

Figure 5-5. The number of eggs laid daily by K28-0-Im virgin females, a partheno-genetic strain established from the sexual Kamuela virgins. Open circles indicate observed points, and a line is drawn to connect these empirical points. The experiment was terminated when one-half of the initial cohort had died. See Templeton et al. (1976a) for details.

conditions. Homozygosity at an X-linked locus for an allele whose frequency in the natural population is about 20% is necessary but not sufficient for the expression of abnormal abdomen. In addition to this major locus, a Y-linked locus and at least three autosomal loci, with the major modifier being the acrocentric II chromosome of *D. mercatorum*, influence the expression of the major locus. The egg-laying capacities of several isofemale lines derived from wild-caught females were examined, and crosses were simultaneously made to determine whether the major allele associated with abnormal abdomen was segregating in the lines. Unfortunately, the assay for the presence of this allele was inefficient at that time, and many of the lines scored as not segregating for the abnormal abdomen should in fact contain this allele. The results are shown in Figure 5-6. The lines not known to be segregating for the abnormal abdomen allele give a bimodal distribution, but the lines known to be segregating at the major locus are unimodal with none in the range of the lower mode of the other lines. Although preliminary, these data imply abnormal abdomen is associated with deleteri-

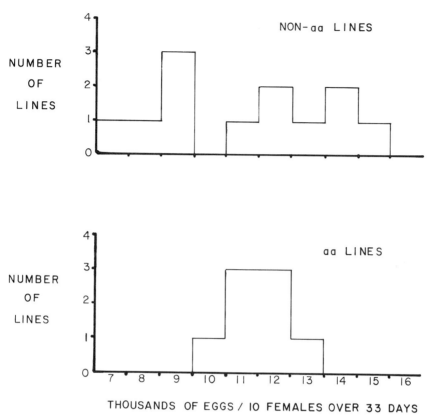

Figure 5-6. The effect of abnormal abdomen (aa) upon the number of eggs produced by the isofemale lines collected from Kamuela in 1976.

ous viability consequences but beneficial fecundity effects. Also, the life history data in Figure 5-5 came from a parthenogenetic line fixed for abnormal abdomen, but in a genetic background that suppressed most of its gross morphological effects.

However, selection on the basis of increased fecundity is obviously not the only aspect of selection engendered by the initial utilization of the parthenogenetic capacity. For example, assuming a homogeneous probability in the sexual ancestors of an unfertilized egg developing, the data in Figure 5-6 imply the frequency of the abnormal abdomen allele should increase from about 20% to 30%. Yet, about two-thirds of the parthenogenetic lines established from the natural population are fixed for the abnormal abdomen allele. Indeed, a more detailed examination of the data indicates that tychoparthenogenetic rates are not homogeneous throughout the ancestral sexual species, implying the existence of genetic differences among the sexual females in tychoparthenogenetic capacity. Carson (1961, 1967a) and Futch (1962) have reported different tychoparthenogenetic capacities in sexual stocks of different geographical origin in three species of *Drosophila*. Templeton (1979a) has shown such heterogeneity can exist even between local populations less than 3 km apart. Moreover, variability occurs within a single geographical population such that a few isofemale lines have rates 10-100 times that of the remainder of the lines established from the same locality (Templeton 1979a, Templeton et al. 1976a), although all initial rates are still quite low (less than 10^{-4}).

One possible basis for these differences was uncovered by Counce and Ruddle (1969) with studies on strain differences in egg structure in *D. hydei*. They found that the chances for the formation of fusion and/or cleavage nuclei in unfertilized eggs depends upon the egg structure, and they concluded that these differences in egg structure could affect the ability of the strains to evolve parthenogenesis. Hence, the chances of evolving a thelytokous population are undoubtedly severely constrained by the genetic variability present in the sexual population that fortuitously affects the chances for parthenogenetic development in unfertilized eggs. Recurring episodes that force females to retain virginity will select upon this variability and will cause an immediate divergence from the sexual ancestors in those characters such as egg structure that can influence parthenogenetic development. One implication of this selection is that there is an immediate increase in parthenogenetic rates even after just one parthenogenetic generation, and observed single-generation increases have been as high as 1000-fold (Templeton et al. 1976a).

The Initial Thelytokous Generation

Given that a virgin female from the sexual population has produced an egg that initiates parthenogenetic development, additional selective events occur that can influence greatly the ultimate evolutionary fate of the thelytokous population. The first such event is a viability-selective bottleneck that is encountered if the mode of parthenogenesis greatly alters the genetic environment from the sexual ancestral state (as in gamete duplication or terminal fusion). Such a bottleneck appears to be a major barrier to the establishment of thelytoky in *D. mercatorum* (Templeton et al. 1976a), which produces primarily via gamete duplication. The natural populations of this species display levels of polymorphism and individual heterozygosity typical for *Drosophila* (Clark et al. 1981, Templeton 1979a) so the total homozygosity that is immediately enforced by gamete duplication represents a large change in genetic environment (Templeton 1979b). All recessive lethals have to be eliminated, and complexes of genes that interact well together under total homozygosity must be selected (Templeton 1979b). Moreover, the genetic changes induced by the selection for fecundity or tychoparthenogenesis in the original sexual females can induce additional selective pressures at this stage. For example, fixation of the major abnormal abdomen allele induces strong selective pressures at its modifier loci (Templeton 1980). Thus, fixation at one major locus can have cascading effects throughout the genome. In addition, parthenogenesis can be regarded as the ultimate founder event (Templeton 1979b), which automatically induces random fixations and strong disequilibria between nonalleles scattered throughout the genome. For example, the parthenogenetic line fixed for abnormal abdomen discussed earlier was also fixed for the X-chromosomal element and acrocentric II autosomal element that drastically reduces female sexual receptivity, as well as an X-chromosomal element and another acrocentric II element that induces male sterility.

Another behavioral trait with important life history implications also seems to fit this pattern of inheritance. The females from this parthenogenetic stock are extremely lethargic in their culture bottles, to the extent that the bottle can be opened and vigorously shaken with little chance of losing a single fly. Their behavior is so marked in this regard, that Dr. Hampton Carson began calling this thelytokous line the "lazy girls." However, in some experiments performed by Dr. J. Spencer Johnston, the "lazy girls" were placed in a wind tunnel with no food and a gentle breeze. After a few minutes, the flies began to groom vigorously and then moved upwind farther than any stock of *mercatorum* examined at the time. Hence, their laziness is conditional. As long as they are on food, they remain on it; as soon as the food is gone, they become strong dispersers. All of these phenotypes are separable in a genetic sense, so they are not pleiotropic manifestations of a single genetic syndrome. Moreover, none of them is identical to abnormal abdomen, which, it will be recalled, has its major locus on the X chromosome and its major modifier on the acrocentric II autosome. However, I doubt that this genetic parallelism results from chance alone. From what is known of abnormal abdomen, very strong selection on the X and acrocentric II would be expected, making hitchhiking effects involving those chromosomes particularly likely. Other parthenogenetic lines fixed for abnormal abdomen do not display all these other traits, but each line has its own unique behavioral, life history, and morphological peculiarities that easily can be explained by hitchhiking effects and random fixations. (Note that the uncovering of these traits in the parthenogenetic lines also has revealed that the ancestral sexual population is very polymorphic for such traits.) These hitchhiking and drift effects in turn may induce further selective alterations and, very importantly, set constraints upon the emerging thelytokous population that determine the ecological conditions under which they may persist and thereby influence all further evolutionary modifications.

The observation in *mercatorum* that an intense viability bottleneck is encountered during the early parthenogenetic generations (Annest and Templeton 1978, Templeton et al. 1976a) and that the parthenogenetic flies can respond successfully to this selection runs afoul of some additional prophecies of parthenogenesis. First, there is the "prophecy of the evolutionary dead end," that parthenogenetic populations simply cannot respond to intense selection because they lack genetic variability. Such a situation certainly does not obtain during the early parthenogenetic generations, even if the entire population is founded by a single virgin female. Automixis insures that the parthenogenetic progeny of even a single virgin from the sexual population will be genetically heterogeneous because of the actions of segregation, assortment, and recombination. The degree of genetic heterogeneity can be very extensive if the founding virgin is highly heterozygous, thus providing the variability that makes effective selection possible.

The empirical studies on the viability bottleneck contradict another prophecy of parthenogenesis. Cuellar (1977) has hypothesized that inbreeding should preadapt a population to parthenogenetic modes that promote rapid homozygosity. However, in *D. mercatorum*, the observed chances for producing a totally homozygous parthenogen increase as the level of heterozygosity increases in the sexual founders both between (Templeton 1979a) and within populations (Templeton et al. 1976a). Moreover, it is easier to obtain totally homozygous parthenogens from outcrossed wild stocks than from inbred laboratory stocks (Templeton et al. 1976a). These results are just the opposite from those expected under an inbreeding preadaptation hypothesis. An explanation for this apparent paradox is provided by some coadaptation experiments

(Templeton 1979b) which show that active selection for a gene complex yielding a viable developmental history under total homozygosity is quantitatively as or more important than selection against recessive lethals during the initial reproductive transition. Inbreeding will reduce the incidence of recessive lethals and lower levels of average heterozygosity, but it cannot be equated with near-total homozygosity as studies on flies (Averhoff and Richardson 1974, 1976, Franklin 1977), chickens (David et al. 1979), and humans (Degos et al. 1974) all indicate. However, by reducing the average heterozygosity of individuals, inbreeding can greatly reduce the amount of interclonal variability produced under automixis during the critical initial generation. This in turn reduces the effectiveness of selection for a coadapted gene complex, thereby lowering the chances for successfully passing the viability-selective bottleneck (Templeton et al. 1976a). Hence the failure of Cuellar's prediction illustrates the extreme importance of automixis and genetic variability during the reproductive transition.

An alternative preadaptation hypothesis was suggested by Nur (1971, 1972). He has shown that most obligate thelytokous populations with gamete duplication have evolved from sexual ancestors with males that are cytologically or genetically haploid. Nur (1971) feels heterosis is less important in such populations and that mutants are selected more for their effects under haploidy or effective homozygosity. Maynard Smith (1978) has added also that recessive lethals and deleterious genes are rarer in such haplodiploid species, but in view of the *D. mercatorum* data this effect is of secondary importance. However, the work with *D. mercatorum* does support the view that male haploidy effectively mimics the selective environment associated with total homozygosity. Templeton et al. (1976b) noted that in their coadaptation experiments on parthenogenetic lines the multilocus fitness properties revealed by the X-linked markers in control sexual males were homogeneous with the parthenogenetic female results but inhomogeneous with the results obtained with the males' sexually produced sisters. Hemizygosity therefore did mimic the selective environment associated with total homozygosity, as expected under Nur's (1971) hypothesis.

The kind of viability bottleneck discussed above is, of course, not nearly so pronounced under central fusion and, indeed, may be virtually nonexistent in an organism with small chromosomes in a recombinational sense, such as *Drosophila*. The reason for this is that central fusion in such organisms maintains to a high degree the genetic environment of the virgin founder. This observation coupled with Nur's hypothesis implies that tychoparthenogenesis in a diploid species is more likely to yield a thelytokous population reproducing primarily by central fusion than by gamete duplication or terminal fusion simply because there is one less barrier encountered during the reproductive transition—the barrier of the viability bottleneck, which can be a severe barrier indeed. Therefore, the mode of parthenogenesis that evolves is less a matter of what mode is optimal by some decision-theoretic criterion, but is far more a matter of what type of genetic system and population structure existed in the ancestral sexual species. Moreover, different sexual populations are apparently genetically biased toward a particular mode of parthenogenesis. For example, in *D. mercatorum*, there is a strong bias toward gamete duplication (Carson et al. 1969), whereas in *D. parthenogenetica* there is a bias toward central and terminal fusion (Stalker 1954). This bias can be explained in part by differences in such attributes as egg structure (Counce and

Ruddle 1979) and the behavior of meiotic products in unfertilized eggs (Doane 1969), attributes that undoubedly have not been selected for on the basis of their parthenogenetic implications in the ancestral sexual population. The types of tychoparthenogenetic variants that exist in the sexual population therefore are a major constraint on the type of thelytoky that ultimately evolves.

After viable parthenogens are selected, a second genetically distinct selective bottleneck is encountered when the viable parthenogens themselves reproduce parthenogenetically. Not only does this bottleneck involve different genes from the viability bottleneck, but it also involves different and/or additional genes from those involved in the initial selection for tychoparthenogenetic capacity in the original sexually produced virgin females. This is indicated by the following data: 141 isofemale lines from a natural population of *D. mercatorum* were screened for parthenogenesis, producing a total of 53 viable parthenogenetic adults. However, only seven of these viable adults were themselves capable of reproducing parthenogenetically during a 2-week period after eclosion. The fact that only 13% of the viable adults could reproduce indicates the severity of this second bottleneck. Moreover, 34 of the 53 parthenogens came from eight "hot" isofemale lines. However, only five of the 34 (15%) parthenogenetic adults from the "hot" lines and two out of 19 (11%) from the "nonhot" lines could themselves reproduce parthenogenetically. This difference is not significant. The genes conferring increased tychoparthenogenetic capacity in the bisexual lines therefore do not confer increased capacity or efficiency of parthenogenetic reproduction once the initial reproductive transition has been made.

Isozyme studies have given an intriguing glimpse of the existence of this second bottleneck. We have found some 17 segregating isozyme loci in a natural population of *D. mercatorum*. We have scored 40 of the viable parthenogenetic adults for their isozymes and found that the allele frequencies in this sample of parthenogenetic adults are consistent with a random sample from the sexual gene pool (Clark et al. 1981). However, of these 40 flies, four came from self-sustaining parthenogenetic clones and 36 were flies incapable of reproducing parthenogenetically. Three of the four self-sustaining clones were identical for all 17 isozyme loci and the remaining line differed only at two loci. Of the 36 scored flies that were incapable of giving rise to a self-sustaining clone, only one had this multilocus isozyme phenotype. Although this phenotype consists of homozygosity for the most common alleles, its probability as calculated by multiplying the single locus allele frequencies from the sexual population is only 0.05. Hence, the one in 36 observed in the flies incapable of parthenogenetic reproduction is close to what is expected by chance alone, but the near identity of all four self-sustaining clones is very unlikely by chance alone (probability is less than 0.0006). Moreover, coadaptation experiments on the self-sustaining clones yielded evidence for very strong viability selection, but none of it could be attributed to the isozyme markers (Templeton 1979b). There seems to be little if any selection on the isozyme markers in going through the viability bottleneck, therefore, but strong selection is apparently encountered at this second bottleneck for parthenogenetic reproduction (Clark et al. 1981).

The First Phase of Clonal Selection

Once selection has produced an array of genotypes that are both viable under the genetic conditions imposed by parthenogenesis and capable of parthenogenetic reproduction themselves, clonal selection will occur to enhance both viability and efficiency of parthenogenetic reproduction. A good illustration of this phase of clonal selection is provided by the work of Annest and Templeton (1978) as shown in Figure 5-7. In this experiment, 150 sexually produced females heterozygous for several marker loci were isolated as virgins. The flies reproduced by gamete duplication and produced 233 parthenogenetic flies containing all possible genotypic categories. However, even at this stage, the frequency distribution was highly nonrandom, indicating strong viability selection. Although all of these 233 flies obviously had nonzero viabilities under parthenogenesis, they only produced 157 parthenogenetic progeny for the next generation, once again demonstrating the reality of the second selective bottleneck. These 157 flies produced 513 flies, the largest increase during any one generation and close to the carrying capacity of our culture system, indicating that a second bottleneck had been passed. From the adults at generation 2 on, therefore, only flies with proved nonzero viabilities and nonzero parthenogenetic rates were in the population. Nevertheless, quantitative differences existed among the clones with respect to these attributes, and Figure 5-7 shows the rapid elimination of most of the clones during this phase of clonal competition.

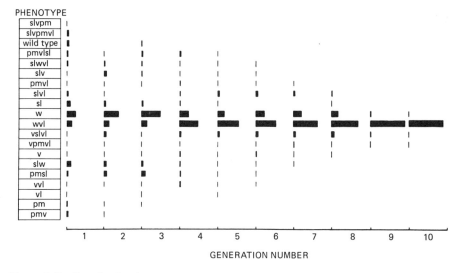

Figure 5-7. Clonal selection among 20 phenotypes over 10 generations of parthenogenetic reproduction in a unisexual, discrete generation population cage. The first-generation females were generated by automictic parthenogenesis in virgin mothers which were heterozygous for six gene markers. All phenotypes in subsequent generations represent one or more completely homozygous genotypes. The length of each bar is proportional to the frequency of each phenotype in each generation. From Annest and Templeton (1978).

Stalker (1954) also found evidence for clonal selection in a unisexual strain of *D. parthenogenetica* that reproduced by a mixture of central and terminal fusion. Stalker found a steady increase in the frequency of eggs initiating development over the first 17 parthenogenetic generations, with the frequency going from 0.91% to 8.2%. Moreover, the survivorship of individuals to the larval stage, given development had initiated, increased from 8.96% to 18.88%. Combining these two factors, the overall rate of parthenogenesis (i.e., the number of viable larvae per number of unfertilized eggs) increased about 20-fold (from 0.08% to 1.55%). Stalker (1956a, b) attributed the lack of response after generation 17 to the exhaustion of the genetic variability carried over by automixis from the original sexual founders. To test this idea, Stalker made some backcrosses at unisexual generation 55 to males from the bisexual ancestral strain followed by additional unisexual and backcross generations. After this influx of new genetic variability, clonal selection was once again effective and the parthenogenetic rate doubled from 1.55% to 3.06%, principally because of the increase in the percentage of eggs initiating parthenogenetic development. These results clearly indicate that a large (20-fold) and rapid response to selection is possible under strictly parthenogenetic reproduction during the early unisexual generations.

A detailed fitness analysis of the results given in Figure 5-7 and similar experiments reveals that clonal selection principally deals with nonadditive and nonmultiplicative fitness interactions between loci (Annest and Templeton 1978). Several studies have shown that selection in sexually reproducing, outcrossing populations works primarily on the additive component of the genotype-phenotype relation, but once outcrossing is restricted, selection can operate very effectively on the total fitness properties of multilocus genotypes (Crow 1957, Malmberg 1977, Moll and Stuber 1971, Templeton 1979b, Templeton et al. 1976b). The reproductive transition therefore is also accompanied by a radical transition in the unit of selection (Templeton 1979b, Templeton et al. 1976b). This means that the response to natural selection in a sexual vs. a parthenogenetic population will be different even if the very same genes, associated with the very same phenotypes subject to the very same type of selection exist in both populations. This will lead to further genetic differentiation between the sexual ancestors and their thelytokous offshoot. This radical fitness restructuring of the genome has been directly observed during the reproductive transition in strains of *Drosophila mercatorum* (Templeton 1979b).

The impact of this change in unit of selection should not be underestimated, for it means there can be an adaptive divergence of the sexual from the parthenogenetic populations even under identical environments. For example, it is well known in quantitative genetics that when a plant or animal population is exposed to an extreme environment that is generally nonoptimal for the species, the resistant phenotype of the survivors is primarily a result of the interaction components rather than the additive components of the genotype-phenotype relationship (Jinks et al. 1973). Therefore, the heritability of tolerance to extreme environments rarely encountered by the bulk of the species is generally close to zero, implying there can be no effective selection for such tolerance. However, once a transition to automixis has taken place, selection can effectively utilize the interaction components so that a tolerant parthenogenetic population may quickly evolve. Such a situation may occur commonly, because one of the factors mentioned earlier that gives the initial impetus to utilizing a tychoparthenogenetic capacity is low density in an ecologically marginal area. The area may well be ecologically marginal because of an extreme environment to which the bulk of the species is rarely exposed. Therefore, the sexual individuals migrating to the margins may not be able to adapt to this extreme environment even if they contain the genetic variants needed to effect such an adaptation. However, clonal selection in a parthenogenetic population could rapidly effect such an adaptation. This change in the unit of selection is probably the primary reason for the oft-noted phenomenon (Lokki et al. 1975, Seiler 1961, 1967, Suomalainen et al. 1976) that many parthenogenetic populations inhabit areas beyond the normal ecological range of their sexual ancestors.

Another change effected by the alteration in the unit of selection is that parthenogenesis can allow the stabilization of a genotype with superior homeostatic capabilities in dealing with fine-grained environmental fluctuations that once again have little or no heritability in the sexual population. As a consequence, Templeton and Rothman (1978) have shown, a newly evolved parthenogenetic population sometimes can evolve such superior homeostatic capabilities that they can drive their sexual ancestors to extinction. Moreover, there is evidence that this situation actually has occurred in the genus *Drosophila*. *Drosophila mangabeirai*, the only natural obligate thelytokous *Drosophila,* has apparently driven its sexual ancestors to extinction. Moreover, it shows very broad geographical and altitudinal (0-4000 ft) ranges for its species group (*willistoni*), thereby indicating a broad ecological tolerance (Carson 1962). Note that the situation illustrated by *D. mangabeirai* and the theory of Templeton and Rothman (1978) represent a violation of one of the more common prophecies of parthenogenesis —that environmental uncertainty favors sexual reproduction.

It is hoped that the above examples demonstrate that the change in the unit of selection that inevitably accompanies the reproductive transition is not a trivial matter; instead, this change may be one of the most important causes of the evolutionary divergence of the parthenogenetic from the sexual populations. It also means that differences in niche or ecological tolerance do not have to evolve slowly after the establishment of the thelytokous population; neither do they require mutational input after separation of the sexual population from the unisexual population. Instead, natural selection causes the genetic variability carried over from the ancestral sexual population to be utilized in ways that are virtually impossible under sexual reproduction. Hence, the divergence between the populations in ecologically and adaptively relevant characters can be extremely rapid during these early generations of clonal selection.

Finally, the very mode of reproduction itself can be altered through clonal selection. First of all, automictic diploids generally retain their capacity to reproduce sexually if they mate. As mentioned earlier, sometimes genetic-based sexual isolation or errors in imprinting provide the initial impetus for parthenogenesis, and in such cases this reprodictive isolation would be expected to persist. Also, other cases involve the geographical isolation of virgin females, and such isolation may persist so that the opportunity to mate is never realized. However, other instances result from a shortage of males. Therefore, the parthenogenetic females are likely once again to have the opportunity of reverting to sexual reproduction.

A great deal of experimental work on the sexual behavior of several species of parthenogenetic *Drosophila* has revealed some interesting facts that are important in this regard. Wolfson (1958) and Wei (1968) both discovered that recently derived unisexual strains have a wide range of sexual receptivity—both higher and lower than that of the bisexual ancestral strains. Nevertheless, their work and additional studies (Carson et al. 1977) reveal a definite bias in favor of sexual isolation, sometimes to the point of virtually complete isolation, even though no attempt has been made deliberately to select for sexual isolation in these studies. What is the explanation for this bias, then? The mating system in *Drosophila* is one of female choice, and this has often led to the evolution of very elaborate courtship rituals and actions that depend upon several behavioral, chemical, and morphological attributes—the so called "mate recognition system" referred to earlier. Paterson (1978) and Templeton (1979c) have argued that such mate recognition systems are normally subject to strong stabilizing selection and are generally polygenic. When the transition to parthenogenesis is made, the females are freed from the constraint of mating to reproduce, and simultaneously the types of selective events discussed earlier occur which obviously involve a great many genes influencing a great many attributes. Either through pleiotropic effects or hitchhiking effects, such an extensive and rapid polygenic alteration will almost always disrupt some aspect of the female's mate recognition system. Because the females now do not need to mate to reproduce, this disruption is not counterbalanced by selection reestablishing a mate recognition system, thus leading to the systematic bias in favor of sexual isolation observed in parthenogenetic *Drosophila*. It should be emphasized that genetically it is far easier to establish sexual isolation in a parthenogenetic form from its sexual ancestors than to establish sexual isolation between two sexual species derived from a common ancestor. In the former case all that is necessary is to destroy the mate recognition system in the absence of any counterbalancing selection, whereas in the latter case the old mate recognition system must be transformed into a functioning new mate recognition system in one or both lines, and yet at all points in time the mate recognition system must still be functional in both incipient species so as to allow mating to occur.

As argued above, sexual isolation tends to arise spontaneously in parthenogenetic populations, but clonal selection could augment this isolation as well. It has been observed that crossing parthenogenetic females to males from sexual strains usually destroys the parthenogenetic capacity in sexual daughters (Carson 1967a, Templeton 1979b). This is not surprising, for it should be recalled that during the reproductive transition the parthenogenetic populations must successfully go through several intense selective events and bottlenecks, which alone constitute a formidable barrier to the utilization of a tychoparthenogenetic capacity (Templeton et al. 1976a). Moreover, successful passage through these selective bottlenecks involves coadapted gene complexes characterized by extremely nonadditive fitness organization. Hence, such complexes are easily destroyed by just one generation of sexual reproduction, causing the parthenogenetic capacity of the daughters to plummet. Any parthenogenetic genotype that did tend to mate would be effectively eliminated from the thelytokous population. Such selection in general would be expected to be quite effective, because as mentioned earlier, the mate recognition system is no longer under stabilizing selection in the thelytokous population, and the associated genetic events accompanying the reproductive transition would amost always release much variability in the mate recognition system, as the empirical studies of Wolfson (1958) and Wei (1968) indicate.

This type of clonal selection for increased reproductive isolation would be particularly effective during the initial generations because of the carryover of genetic variability from the sexual ancestors and its release by automictic reproduction. This prediction is supported by the empirical studies of Doerr (1967) and Ikeda and Carson (1973) on selection for increased reproductive isolation in parthenogenetic populations recently derived from sexually produced females. However, such selection also can be effective in parthenogenetic populations long after the reproductive transition. Henslee (1966) placed males with females from a substrain of parthenogentic *D. mercatorum* that had been established several years before. He then allowed the females that did not mate to reproduce parthenogenetically and tested their parthenogenetic descendants for sexual isolation. He found that the degree of sexual isolation was substantially increased. Thus, effective clonal selection was still possible in this long-established parthenogenetic strain which reproduced almost exclusively by gamete duplication. Henslee's results demonstrate that long-established parthenogenetic populations can respond to selection, contrary to the "prophecy of the evolutionary dead end."

A second type of clonal selection on reproductive biology also occurs in these initial unisexual generations—selection on the mode of automixis itself. All examined cases of tychoparthenogenesis in *Drosophila* have revealed a mixture of automictic mechanisms occurring in the unfertilized eggs, even in the eggs derived from a single female. For example, *D. parthenogenetica* reproduces by a 50:50 mixture of central and terminal fusion (Stalker 1954) and *D. pallidosa* by a mixture of gamete duplication, terminal fusion, and perhaps some central fusion (Futch 1979). This tendency to have several automictic modes simultaneously has been studied most extensively in *D. mercatorum*. This species reproduces by a mixture of gamete duplication and central fusion, although a small amount of terminal fusion cannot be excluded (Templeton 1982). Carson (1973) and Annest (1974) found that the amount of duplication versus fusion is partially under genetic control with the percentage of fusion varying from 1 to 43% depending upon the parthenogenetic strain. Consequently, the very mode of parthenogenesis is genetically variable and subject to clonal selection. This type of selection has been observed directly by Annest (1976). He initiated parthenogenetic population cages with flies characterized by high fusion rates and observed that the high fusion rates were quickly lost in favor of gamete duplication.

It also is reasonably certain that a similar type of clonal selection occurred in *Drosophila mangabeirai*, the only natural obligate thelytokous *Drosophila*. First, *D. mangabeirai* only has central fusion, and in light of the universality of automictic mixtures in tychoparthenogenetic *Drosophila*, this exclusiveness is most likely an evolved condition. Second, the cytological investigations by Murdy and Carson (1959) support this proposition. Figure 5-8A shows the events of meiosis in *D. melanogaster* (and almost all other *Drosophila*) under sexual reproduction. The spindles at both meiotic divisions are oriented perpendicularly to the long axis of the egg, causing the four pronuclei to line up on this perpendicular. The innermost nucleus usually becomes the egg nucleus, and all cleavage divisions apparently must take place in the central cytoplasm. However, in *D. mangabeirai*, there is a strong tendency for the first meiotic division to be parallel to the long axis of the egg, whereas the second division is perpendicular (Figure 5-8B). This places the pronuclei that separated at meiosis I into the inner part of the egg, where they fuse and initiate cleavage divisions. It is hard to believe that this unique configuration has arisen without selection, and specifically a type of selection that only operates after the establishment of exclusive parthenogenetic reproduction.

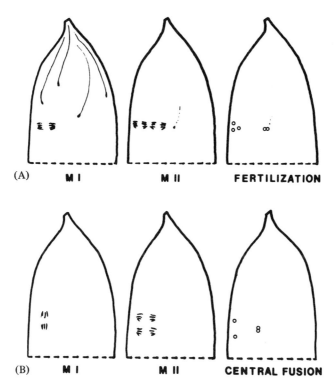

Figure 5-8. Meiosis in *Drosophila melanogaster* under sexual reproduction (A) and in *Drosophila mangabeirai* under central fusion (B). MI depicts a longitudinal section of the meiosis I spindle orientation, MII a longitudinal section of the meiosis II spindle orientation, and finally a longitudinal section of the configuration of the haploid pronuclei either at the moment of fertilization (A) or at central fusion (B). Modified from Murdy and Carson (1959).

These lines of evidence indicate that the early parthenogenetic generations are characterized by selection resulting in the enhanced stabilization of one automictic mode. The resulting automictic mode is determined in part by this selection and in part by the type of genetic variability present in the ancestral sexual population. This selection, coupled with the other selective forces operating during the reproductive transition and initial unisexual generations, illustrates the extreme dynamism of this evolutionary situation. Rapid response to selection at this stage is critical and is possible only because the genetic variability carried over from the sexual ancestors can be released via automixis. This evolutionary flexibility would be lacking if the initial parthenogens were apomictic. Hence, contrary to the prophecies of parthenogenesis, automixis represents a most promising starting point for the origin of a parthenogenetic variety.

The Second Phase of Clonal Selection

The selective forces encountered during these early unisexual generations eventually will exhaust the genetic variability initially released by automixis, as is shown by Figure 5-7. However, it would be a mistake to assume that evolution stopped at this stage. Instead, there begins a second phase of clonal selection operating upon newly arisen mutants. Once again, it is important to realize that these mutants will be selected only for their effects upon a given genetic background and not for some generalized additive or combining abilities. Moreover, Leigh (1970) and Eshel (1973) have shown that clonal selection can modify mutation rates very efficiently so as to increase the expected clone survival probability. Therefore further selective improvement is expected during this second phase of clonal selection. Note that these predictions violate yet another prophecy of parthenogenesis: Muller's ratchet. This prophecy proclaims that under certain conditions asexual populations irreversibly accumulate deleterious mutants. However, it is important to note that this prophecy originally was applied only to asexual haploids and has never been shown even theoretically to apply to automictic diploids. Heller and Maynard Smith (1979) have applied the ratchet to selfing diploids and claim, without evidence, that the ratchet may also be important in parthenogenetic populations. However, as I have discussed elsewhere (Templeton 1982), the application of the ratchet to selfing involves assumptions known to be seriously violated under parthenogenesis and ignores the results of Eshel (1973) and Leigh (1970). It is therefore questionable how much applicability Muller's ratchet has to parthenogenesis.

There is some empirical evidence on this issue. In *D. mercatorum*, some parthenogenetic strains have existed since 1961 (Carson 1967a). I have repeatedly used one of these, S-1-Im, as a control in many of my experiments and hence periodically have measured its parthenogenetic rate (number of adults per number of unfertilized eggs). This rate has not degenerated as expected under a ratchet model (despite the fact that the total population size of this strain has rarely exceeded 1000 and is usually on the order of 50-100 and occasionally has been deliberately reduced to one for several generations—all factors that theoretically make the ratchet more likely to operate). Instead, it has displayed a steady and gradual improvement, going from an initial rate of 2.5% in March 1962 (Carson 1967a) to 6.0% in March 1980 (Templeton, unpublished data). This indicates a pattern of incorporation of beneficial mutants, each of small effect.

Another strain, however, has shown a dramatic and discontinuous increase in parthenogenetic rate. In 1974, I initiated a parthenogenetic strain, K23-0-Im, from one of my screenings of a natural population of *D. mercatorum*. The parthenogenetic rate quickly stabilized at about 0.2% and has remained there since in most subclones of this strain. However, Dr. Charles F. Sing and Robert Clark at the University of Michigan used this strain in some experiments on glycolytic intermediates. As part of their studies, they had isolated in 1979 several parthenogenetic daughters of a single female parent and noticed that in two vials there were many larvae rather than the usual handful. I later studied the strains descended from these two individuals and discovered that both of them have an eye-color mutation that maps to the acrocentric I autosome, and both have rates of parthenogenesis preliminarily estimated at 10%, a 50-fold increase from the original stocks and the maximum recorded for *D. mercatorum*. The detailed genetics of this increase and its possible relation to the eye-color mutation are currently being investigated, but this example clearly illustrates that large gains in parthenogenetic efficiency can be achieved. The fact that this large increase has occurred in only one strain illustrates the dependence on random mutation during this second phase of clonal selection. Most likely, the successful thelytokous populations in nature have had such a beneficial mutation early so that much of the cost of parthenogenesis in terms of failure of egg development was eliminated, perhaps occasionally to the point where a "cost of sex" seems to arise. However, by the time parthenogenetic efficiency has reached this point, the thelytokous and bisexual populations would have diverged so greatly for reasons previously discussed that any contrast made only on the basis of their abilities to lay viable female-producing eggs would be of dubious ecological or evolutionary validity.

A second major tendency occurring during this second phase of clonal selection is for automixis to evolve into apomixis. As evolution progresses in any automictic clone, the fitness properties of the clone become increasingly dependent upon maintaining a coadapted, multilocus gene complex. This inevitably leads to increasing selection for any attribute that aids in keeping the gene complex intact. These selective forces lead to particularly interesting results if the automictic mode is central fusion. This mode is very important, for as explained earlier, central fusion mitigates or even eliminates the viability bottleneck. For example, a diploid sexual species having tychoparthenogenesis with much initial central fusion is more likely to give rise to thelytokous offshoots than sexual species with little central fusion. Templeton (1974) has shown that if an initial heterosis is present, or if a multilocus heterotic effect evolves, selection favors increasing amounts of central fusion. Moreover, the pressures are self-reinforcing, so there is greater and greater selection for increased amounts of central fusion. Thus, this model predicts that if the right type of tychoparthenogenetic variants exist in the sexual ancestors, there would be a strong tendency for central fusion to become the sole automictic mode. Once central fusion is established, there is also selection favoring a reduction in the probability of recombination with the centromere, since those clones that can reduce this probability are more able to maintain their gene complexes intact. The fate of the genomes under central fusion has been discussed in more detail by Asher (1970). He argues that selection will favor (1) lower efficiency of synapsis, (2) lower degree of interference, (3) production of smaller chromosomes, and (4) production of chromosomal rearrangements suppressing crossing over. Work with *Drosophila* indicates that genetic variants for (1) and (4) commonly occur, so perhaps routes (1) and (4) represent the most common evolutionary fates of genomes under central fusion. A possible example of route (4) is provided by *D. mangabeirai*, which has inversions close to the centromere that suppress crossing over for much of the genome (Templeton 1974, Carson 1967b). A similar example of probable karyotypic evolution after the establishment of central fusion is provided by the fly *Lonchoptera dubia* as discussed in Ochman et al. (1980). An example of route (1) is provided by the work of Nur (1979) on a parthenogenetic coccid in which synapsis is completely suppressed followed by a normal meiosis and central fusion. These models explain well the observed fact that among naturally occurring obligate automictic thelytokous organisms derived from diploid sexual ancestors, the most common type of parthenogenesis is central fusion coupled with some form of partial or complete crossover suppression.

It should be recalled that apomixis is parthenogenetic reproduction by the suppression of meiosis, and it should be noted that the above-discussed evolutionary tendencies suppress to some extent the effects of meiosis. Indeed, debates over whether such cases as the coccid described by Nur (1979) are automictic or apomictic motivated Nur (1979) to develop a new terminology that avoids the ambiguity of apomixis and automixis. However, this ambiguity, as well as the previously noted results that apomictic insects most likely had automictic ancestors (Lokki and Saura 1980, Seiler 1947, Stalker 1956c), is quite consistent with the prediction of this theory. In effect, the change in the unit of selection automatically induces a successional process under which central fusion tends to evolve into apomixis. This, then, is an evolutionary answer for Maynard Smith's observation on the commonness of apomixis in the insects.

Figure 5-9 summarizes my arguments in terms of an evolutionary flow sheet diagramming the events leading to the establishment of an obligate thelytokous population. As is evident from a glance at this figure, this scheme certainly lacks the simplicity that a prophecy of parthenogenesis provides, but the flow sheet takes into account the genetic and developmental constraints concerning parthenogenesis that are needed to translate accurately evolutionary theory into specific predictions. However, this flow sheet shows that many predictions of the prophecies of parthenogenesis are met, although cause and effect are reversed. For example, the prophecy that "weedy" or "r-type" environments select for parthenogenesis is consistent with the predictions of this flow sheet, but instead of r selection favoring parthenogenesis, parthenogenesis favors the production of "r-selected" phenotypes and induces changes in the unit of selection that allow more effective adaptation to disturbed or extreme environments than in the sexual ancestors. Hence, the evolution of parthenogenesis often leads to the exploitation of a "weedy" or disturbed environment rather than a "weedy" environment leading to the evolution of parthenogenesis. In this regard, the prophecies of parthenogenesis share a fundamental characteristic with the prophecy of a virgin birth that I mentioned at the beginning of this chapter. The widespread belief that a prophecy of a virgin birth can be found in Isaiah owes its existence to the account in Matthew rather than the account in Matthew owing its existence to the actual occurrence of such a prophecy in Isaiah. And so it is with many other prophecies of parthenogenesis.

Summary

One of the most fundamental life history attributes of any organism is its mode of reproduction. Recently, there has been much interest concerning the significance of sexual vs. asexual and parthenogenetic reproduction. I have tested some of the hypotheses in this area with populations of *Drosophila mercatorum*. *Drosophila mercatorum* is normally a sexually reproducing fly whose females are capable of parthenogenetic reproduction if isolated as virgins. Meiosis is retained during parthenogenesis, and diploidy is restored by the fusion of two haploid cleavage nuclei. Because parthenogenetic females are diploid and produce eggs retaining meiosis, it is easy to reestablish

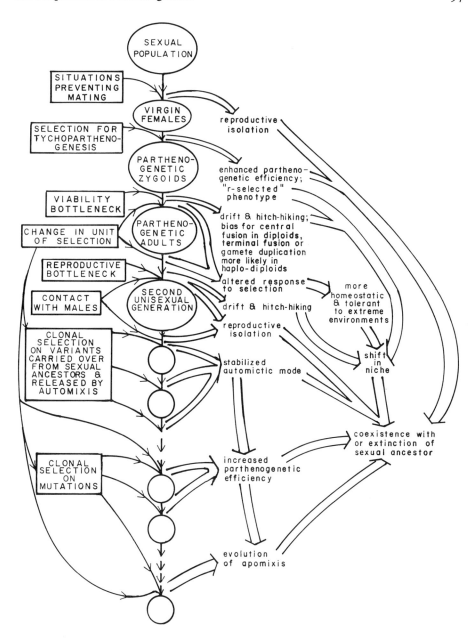

Figure 5-9. A flow sheet diagramming the events leading to the establishment of a parthenogenetic population from sexual ancestors through diploid automixis. Populations are indicated by circles or ellipses, selective forces are in capital letters and boxed, and the effects of the selective forces are in small letters.

sexual reproduction in most cases. Studies with this sexual/parthenogenetic system and other parthenogentic *Drosophila* have revealed that such concepts as "the cost of meiosis," "Muller's ratchet," etc., are irrelevant in the establishment of a parthenogenetic population from sexual ancestors. Also, such an establishment is not an event (as it is usually portrayed) but is rather an extremely dynamic evolutionary process that is associated with many basic life history alterations. It is the nature of these associated life history changes, rather than parthenogenesis per se, that explains the successful establishment of a parthenogenetic population, although it is important to note that certain life history phenotypes are developmentally or physiologically incompatible with normal sexual reproduction and therefore require parthenogenesis or some other asexual mode of reproduction for their expression.

Acknowledgments. This work was supported in part by Grant DEB 78-10455 from the National Science Foundation and Grants 1 R01 GM 27021-01 and 1 R01 AG0 2246-01 from the National Institutes of Health. This paper is dedicated to Dr. Harrison Stalker and Dr. Hampton Carson, whose efforts laid an excellent foundation for all subsequent work on parthenogenetic *Drosophila* and who have been a continual source of ideas, encouragement, and support for my own research in this area.

References

Annest, J. L.: Nuclear fusion in an outcrossed parthenogenetic strain of *Drosophila mercatorum*. Master's thesis, Dept. of Genetics, University of Hawaii, Honolulu, 1974.

Annest, J. L.: Genetic response in unisexual and bisexual laboratory populations of *Drosophila mercatorum*. Ph.D. dissertation, Dept. of Genetics, University of Hawaii, Honolulu, 1976.

Annest, J. L., Templeton, A. R.: Genetic recombination and clonal selection in *Drosophila mercatorum*. Genetics 89, 193-210 (1978).

Argyle, A. W.: The Gospel According to Matthew. Cambridge: Cambridge University Press, 1963.

Asher, J. H., Jr.: Parthenogenesis and genetic variability. Ph.D. dissertation, Dept. of Zoology, University of Michigan, Ann Arbor, 1970.

Averhoff, W. W., Richardson, R. H.: Pheromonal control of mating patterns in *Drosophila melanogaster*. Behav. Genet. 4, 207-225 (1974).

Averhoff, W. W., Richardson, R. H.: Multiple pheromone system controlling mating in *Drosophila melanogaster*. Proc. Natl. Acad. Sci. (U.S.) 73, 591-593 (1976).

Carson, H. L.: Rare parthenogenesis in *Drosophila robusta*. Am. Nat. 95, 81-86 (1961).

Carson, H. L.: Fixed heterozygosity in a parthenogenetic species of *Drosophila*. Univ. Texas Publ. No. 6205, Studies in Genetics 2, 55-62 (1962).

Carson, H. L.: Selection for parthenogenesis in *Drosophila mercatorum*. Genetics 55, 157-171 (1967 a).

Carson, H. L.: Permanent heterozygosity. Evol. Biol. 1, 143-168 (1967 b).

Carson, H. L.: The genetic system in parthenogenetic strains of *Drosophila mercatorum*. Proc. Natl. Acad. Sci. (U.S.) 70, 1772-1774 (1973).

Carson, H. L., Teramoto, L. T., Templeton, A. R.: Behavioral differences among isogenic strains of *Drosophila mercatorum*. Behav. Genet. 7, 189-197 (1977).

Carson, H. L., Wei, I. Y., Niederkorn, J. A., Jr.: Isogenicity in parthenogenetic strains of *Drosophila mercatorum*. Genetics 63, 619-628 (1969).

Carson, H. L., Wheeler, M. R., Heed, W. B.: A parthenogenetic strain of *Drosophila mangabeirai* Malogolowkin. Univ. Texas Publ. No. 5721, Genetics of *Drosophila* 115-122 (1957).

Clark, R. L., Templeton, A. R., Sing, C. F.: Studies of enzyme polymorphisms in the Kamuela population of *D. mercatorum*. I. Estimation of the level of polymorphism. Genetics 98, 597-611 (1981).

Counce, S. J., Ruddle, N. H.: Strain differences in egg structure in *Drosophila hydei*. Genetica 40, 324-338 (1969).

Crow, J. F.: Genetics of DDT resistance in *Drosophila*. Cytologia Proc. Int. Genet. Symp. 1956 (Suppl.), 408-409 (1957).

David, C. S., Kaeberle, M. L., Nordskog, A. W.: Genetic control of immunoglobulin allotypes in the fowl. Biochem. Genet. 3, 197-207 (1969).

Degos, L., Colombani, J., Chaventre, A., Bengtson, B., Jacquard, A.: Selective pressure on HL-A polymorphism. Science 249, 62-63 (1974).

Doane, W. W.: Completion of meiosis in uninseminated eggs of *Drosophila melanogaster*. Science 132, 677-678 (1969).

Doerr, C. A.: Artificial selection for sexual isolation within a species. Master's Thesis, Dept. of Zoology, Washington University, St. Louis, Mo, 1967.

Eshel, I.: Clone selection and the evolution of modifying features. Theor. Pop. Biol. 4, 196-208 (1973).

Franklin, I. R.: The distribution of the proportion of the genome which is homozygous by descent in inbred individuals. Theor. Pop. Biol. 11, 60-80 (1977).

Futch, D. G.: Hybridization tests within the *cardini* species group of the genus *Drosophila*. Univ. Texas Publ. No. 6205, Studies in Genetics 2, 539-554 (1962).

Futch, D. G.: Parthenogenesis in Samoan *Drosophila ananassae* and *Drosophila pallidosa*. Genetics 74, s86-s87 (1973).

Futch, D. G.: Intra-ovum nuclear events proposed for parthenogenetic strains of *Drosophila pallidosa*. Genetics 91, s36-s37 (1979).

Heller, R., Maynard Smith, J.: Does Muller's ratchet work with selfing? Genet. Res. 32, 289-293 (1979).

Henslee, E. D.: Sexual isolation in a parthenogenetic strain of *Drosophila mercatorum*. Am. Nat. 100, 191-197 (1966).

Ikeda, H., Carson, H. L.: Selection for mating reluctance in females of a diploid parthenogenetic strain of *Drosophila mercatorum*. Genetics 75, 541-555 (1973).

Jinks, J. L., Perkins, J. M., Pooni, H. S.: The incidence of epistasis in normal and extreme environments. Heredity 31, 263-270 (1973).

Lamb, R. Y., Willey, R. B.: Are parthenogenetic and related bisexual insects equal in fertility? Evolution 33, 771-774 (1979).

Leigh, E. G., Jr.: Natural selection and mutability. Am. Nat. 104, 301-306 (1970).

Lokki, J., Saura, A.: Polyploidy in insect evolution. In: Polyploidy: Biological Relevance. Lewis, W. (ed.). New York: Plenum Press, 1980.

Lokki, J., Suomalainen, E., Saura, A., Lankinen, P.: Genetic polymorphism and evolution in parthenogenetic animals. II. Diploid and polyploid *Solenobia triquetrella* (Lepidoptera: Psychidae). Genetics 79, 513-525 (1975).

Malmberg, R. L.: The evolution of epistasis and the advantage of recombination in populations of bacteriophage T4. Genetics 86, 607-621 (1977).

Maynard Smith, J.: The Evolution of Sex. Cambridge: Cambridge University Press, 1978.

Moll, R. H., Stuber, C. W.: Comparisons of response to alternative selection procedures initiated with two populations of maize (*Zea mays* L.). Crop Sci. 11, 706-711 (1971).

Murdy, W. H., Carson, H. L.: Parthenogenesis in *Drosophila mangabeirai* Malog. Am. Nat. 93, 355-363 (1959).

Nur, U.: Parthenogenesis in Coccids (Homoptera). Am. Zool. 11, 301-308 (1971).

Nur, U.: Diploid arrhenotoky and automictic thelytoky in soft scale insects (Lecaniidae: Coccoidea: Homoptera). Chromosoma 39, 381-402 (1972).

Nur, U.: Gonoid thelytoky in soft scale insects (Coccidae: Homoptera). Chromosoma 72, 89-104 (1979).

Ochman, H., Stille, B., Niklasson, M., Selander, R. K., Templeton, A. R.: Evolution of clonal diversity in the parthenogenetic fly *Lonchoptera dubia*. Evolution 34, 539-547 (1980).

Paterson, H. E. H.: More evidence against speciation by reinforcement. S. Afr. J. Sci. 74, 369-371 (1978).

Pruzan, A., Ehrman, L., Perelle, I., Probber, J.: Sexual selection, *Drosophila* age and experience. Experientia 35, 1023-1025 (1979).

Seiler, J.: Die Zytologie eines parthenogenetischen Rüsselkäfers, *Otiorrhynchus sulcatus* F. Chromosoma 3, 88-109 (1947).

Seiler, J.: Untersuchungen über die Entstehung der Parthenogenese bei *Solenobia triquetrella* F.R. (Lepidoptera, Psychidae). III. Die geographishe verbreitung der drei Rassen von *Solenobia triquetrella* (bisexuell, diploid und tetraploid parthenogenetisch) in der schweiz und in den angren zenden Lädern und die Deziehung Zur Eiszeit. Z. Vererbag 2, 261-316 (1961).

Seiler, J.: Untersuchungen über die Entstehung der Parthenogenese bei *Solenobia triquetrella* F.R. (Lepidoptera, Psychidae). VII. Versuch einer experimentellen Analyse der Genetik der Parthenogenese. Mol. Gen. Genet. 99, 274-310 (1967).

Sene, F. M.: Effect of social isolation on behavior of *D. silvestris* from Hawaii. Proc. Hawaii Entomol. Soc. 22, 469-474 (1977).

Stalker, H. D.: Diploid parthenogenesis in the *cardini* species group of *Drosophila*. Genetics 36, 577 (1951).

Stalker, H. D.: Diploid and triploid parthenogenesis in the *Drosophila cardini* species group. Genetics 37, 628-629 (1952).

Stalker, H. D.: Parthenogenesis in *Drosophila*. Genetics 39, 4-34 (1954).

Stalker, H. D.: On the evolution of parthenogenesis in *Lonchoptera* (Diptera). Evolution 10, 345-359 (1956 a).

Stalker, H. D.: Selection within a unisexual strain of *Drosophila*. Genetics 41, 662 (1956 b).

Stalker, H. D.: A case of polyploidy in Diptera. Proc. Natl. Acad. Sci. (U.S.) 42, 194-199 (1956 c).

Suomalainen, E., Saura, A., Lokki, J.: Evolution of parthenogenetic insects. Evol. Biol. 9, 209-257 (1976).

Templeton, A. R.: Density dependent selection in parthenogenetic and self-mating populations. Theor. Pop. Biol. 5, 229-250 (1974).

Templeton, A. R.: The parthenogenetic capacities and genetic structures of sympatric populations of *Drosophila mercatorum* and *Drosophila hydei*. Genetics 92, 1283-1293 (1979 a).

Templeton, A. R.: The unit of selection in *Drosophila mercatorum*. II. Genetic revolutions and the origin of coadapted genomes. Genetics 92, 1265-1282 (1979 b).

Templeton, A. R.: Once again, why 300 species of Hawaiian *Drosophila*? Evolution 33, 513-517 (1979 c).

Templeton, A. R.: The theory of speciation via the founder principle. Genetics 94, 1011-1038 (1980).

Templeton, A. R.: Natural and experimental parthenogenesis. In: The Genetics and Biology of *Drosophila*. Vol. 3. Ashburner, M., Carson, H. L., Thompson, N. J., Jr. (eds.). New York: Academic Press, 1982, in press.

Templeton, A. R., Rothman, E. D.: Evolution and fine-grained environmental runs. In: Foundations and Applications of Decision Theory. Vol. II. Hooker, C. A., Leach, J. J., McClennen, E. F. (eds.). Dordrecht, Holland: Reidel, 1978, pp. 131-183.

Templeton, A. R., Carson, H. L., Sing, C. F.: The population genetics of parthenogenetic strains of *Drosophila mercatorum*. II. The capacity for parthenogenesis in a natural, bisexual population. Genetics 82, 527-542 (1976 a).

Templeton, A. R., Sing, C. F., Brokaw, B.: The unit of selection in *Drosophila mercatorum*. I. The interaction of selection and meiosis in parthenogenetic strains. Genetics 82, 349-376 (1976 b).

Wei, I.: Mode of inheritance and sexual behavior in the parthenogenetic strains of *Drosophila mercatorum*. Master's thesis, Dept. of Zoology, Washington University, St. Louis, Mo., 1968.

Wolfson, M.: A study of unisexual and bisexual strains of *Drosophila parthenogenetica* with particular attention to a new microscporidian infection. Master's thesis, Washington University, St. Louis, Mo., 1958.

Chapter 6

Competition and Adaptation Among Diploid and Polyploid Clones of Unisexual Fishes

R. Jack Schultz*

Introduction

During the development of population biology as a science, genetic variation has been regarded as adaptive, maladaptive, or simply waiting to be drawn upon during times of environmental change. Attempts to evaluate the importance of variation, either in the production of heterosis or as adaptive polymorphisms in homogeneous, heterogeneous, or fluctuating environments, are generally confounded by two major problems: (1) identifying the forces of selection that are operational and (2) determining the phenotype (and associated genotype) upon which selection operates. In sexually reproducing organisms the effects of selection are continually obscured by recombination. Asexual organisms provide a simpler system for investigating the adaptive significance of genetic variation. Here, the genetic variation of a population is contained in the form of multiple genetic clones that, in the absence of recombination, magnify the effects of selection in successive generations.

Viviparous fishes of the genus *Poeciliopsis* are especially suited to studying variation because they provide biotypes that reproduce sexually, semisexually, and asexually. All of these forms, which live in desert streams in northwestern Mexico, are sympatric, and in nature presumably compete with each other. In some streams, little more than a meter wide and a few centimeters deep, if all of the species, clones, and hemiclones are counted, as many as 13 biotypes of *Poeciliopsis* coexist.

The origin of unisexuality in *Poeciliopsis* has been traced to hybridization between *P. monacha* (Miller) and *P. lucida* (Miller), two bisexual species (or more accurately, gonochoristic species). The result has been the production of a diploid all-female "species" that now lives with *P. lucida,* upon which it depends for sperm. This hybrid, called *P. monacha-lucida* (Schultz 1969), has spread throughout most of the Rios Mocorito, Sinaloa, and Fuerte, at many localities surpassing its sexual host species in abundance (Schultz 1977).

The reproductive mechanism of *P. monacha-lucida,* called hybridogenesis (Schultz 1969), perpetuates its all-female nature as well as preserving the original proportions of

*Biological Sciences Group, University of Connecticut, Storrs, Connecticut 06268 U.S.A.

the parent's genetic contributions. During or prior to meiosis in the hybrid, the entire paternal (lucida) genome is eliminated so that only the maternal (monacha) chromosomes are transmitted to the eggs of *P. monacha-lucida* (Cimino 1972, Schultz 1961, 1966, Vrijenhoek 1972); with only half of the genes inherited clonally, the term "hemiclone" seems most appropriate for these fish. Each generation, then, combines a clonally inherited monacha genome with a lucida genome drawn anew from the *P. lucida* gene pool. *Poeciliopsis monacha-lucida,* therefore, is a perpetual F_1 hybrid.

This diploid unisexual apparently served as a stepping stone to the development of two triploid forms through the addition of another genome to the monacha-lucida stem, thus producing *P. monacha-2 lucida* and *P. 2 monacha-lucida* (Schultz 1967, 1977). The triploids also produce only female offspring but this time by gynogenesis, a totally clonal form of inheritance wherein sperm from males of *P. lucida* or *P. monacha* trigger embryogenesis in triploid eggs but do not contribute genes to the 3n offspring.

Through the use of electrophoresis (Vrijenhoek 1978, Vrijenhoek et al. 1978) and tissue graft analysis (Angus and Schultz 1979, Moore 1977) multiple clones and hemiclones of the gynogenetic and hybridogenetic fishes have been identified: (1) *P. monacha-2 lucida* has two clones; (2) *P. 2 monacha-lucida*, three; and (3) *P. monacha-lucida*, 18 hemiclones. Some clones and hemiclones are abundant, others less so; some are widespread, others are restricted (Figure 6-1).

With so many closely related biotypes coexisting within a single community it is reasonable to ask whether the clones and hemiclones identified by electrophoresis and tissue graft analyses are merely recognizable genotypes or whether they are really biological entities, drawing differentially on community resources. Does heterosis in these unisexual hybrids contribute to their success relative to parental forms and does their polyclonal structure provide niche breadth?

In the desert habitats of these fishes intense forces of selection are available that have the potential to operate differently on the various biotypes. Among the most obvious physical parameters are changes in water supply and changes in temperature, both daily and seasonally. A pronounced dry season extending from October to June reduces many of the tributaries of major rivers to a series of isolated pools that decrease in size as the season progresses. Thousands of individuals are compressed into small pools where space and food become increasingly limited and water quality deteriorates. When the rains come, the survivors are released and the "race" to rebuild begins. Two major categories of adaptation are likely to emerge under these fluctuating conditions: (1) the capacity to endure intensive competition for food and space under conditions of stress and (2) reproductive systems capable either of functioning under stress or of responding rapidly to favorable conditions. Each of these categories breaks down into a myriad of specific adaptations which by elaborate means we hope eventually to separate and assess; for now, however, let us answer two simple and direct questions: (1) Are genetically recognizable clones and hemiclones biologically different; (2) Does heterosis contribute to the fitness of the unisexual hybrids relative to their parental precursors?

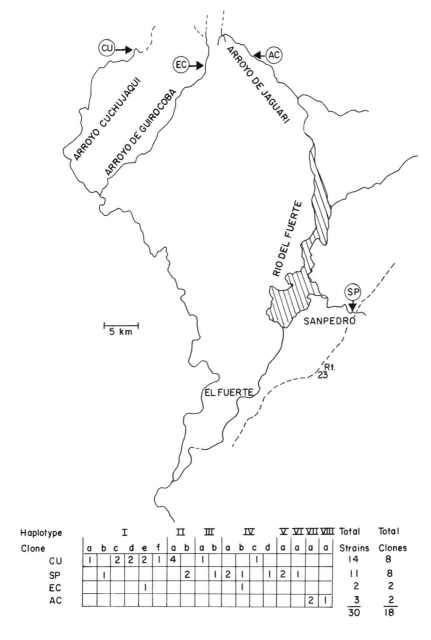

Haplotype	I						II		III		IV				V	VI	VII	VIII	Total Strains	Total Clones
Clone	a	b	c	d	e	f	a	b	a	b	a	b	c	d	a	a	a	a		
CU	1		2	2	2	1	4		1					1					14	8
SP		1					2		1	2	1				1	2	1		11	8
EC				1									1						2	2
AC																	2	1	3	2
																			30	18

Figure 6-1. Collection sites and distribution of hemiclones of *Poeciliopsis monacha-lucida* from the Rio Fuerte in northwestern Mexico. The Roman numerals represent different electrophoretic types detected in the monacha halves of the hybrids (haplotypes). The lower case letters, under each haplotype, indicate clones (actually hemiclones) identified by tissue graft analysis. "Strain" refers to a laboratory line initiated from a single wild female. Collection sites are abbreviated as follows: Arroyo Cuchujaqui, Cu; Arroyo de Guirocoba at El Cajon, EC; Arroyo de Jaguari at Agua Caliente, AC; and Rio San Pedro, SP. From Angus and Schultz (1978).

Peculiarities of *Poeciliopsis* as Research Material

The multiple origins and reproductive processes in unisexual forms of *Poeciliopsis* provide not only the capacity to compare adaptations among clones but also a means of comparing genomes. When a *P. monacha* female hybridizes with a *P. lucida* male, several hundred new hemiclones may be produced from this single fertilization. Whatever genetic recombinants are contained in the monacha eggs are captured in the monacha-lucida system, where they are held intact, protected from recombination. The fact that *P. monacha-lucida* can be synthesized in the laboratory by mating *P. monacha* to *P. lucida* (Schultz 1973) suggests that the process is still going on in those few headwater tributaries where the paternal species overlap. Even though the initial cross does not go especially well because of differences in egg-embryo sizes, in the course of time many thousands of monacha genomes may have entered unisexual life. Competition with the parental species, with other fishes in the community, and with established clones and hemiclones, as well as selection by density-independent factors, are all likely to determine which genomes survive (Schultz 1971, 1973, McKay 1971, Moore 1976, Thibault 1978).

By combining the monacha genomes from successful hemiclones with a standardized lucida genome, differential fitness components among them (e.g., growth rate, fecundity, resistance to stress) can be studied. The standardized *P. lucida* line used in these studies is one that was initiated in 1961 and since then has been through more than 40 generations of brother-sister matings. It is sufficiently inbred that the 23 electrophoretic loci tested (Vrijenhoek et al. 1978) are all homozygous, and tissue grafts (heart, liver, spleen, and scales) from one individual into the musculature of another are successful (Angus and Schultz 1979). Any differences in graft response between two hemiclones of *P. monacha-lucida*, bearing the same lucida genome are therefore a test of the differences between monacha genomes.

Production of exclusively female progeny by hybrid "species" of *Poeciliopsis* has been stressed in earlier papers as a trait that provides a high fitness value (Schultz 1971, Moore 1976); it will not be discussed further here.

Growth Rate Experiments

One of the most important facets of competition in these fishes is growth rate. Under aquarium conditions, if a young fish dies prior to reaching maturity, it is usually the smallest one in the tank. In nature attainment of a moderate to large size removes young from much of the insect predation to which they are subject, as well as from other environmental hazards; it enables them to range more widely for food, to be more aggressive in defense of space, and to produce large broods of young sooner.

To establish whether *P. monacha-lucida* grows faster than *P. lucida*, its sexual host species, broods of young from each biotype ranging in number from 1 to 5 and born within the same 48-hr period were reared together in the same tank. After about 3 months, the total lengths of females were compared. Because the growth rate of males may be different from females, and because males almost completely stop growing after reaching sexual maturity, they were excluded from the analysis. In 13 trials involving seven different hemiclones of *P. monacha-lucida*, the unisexuals grew larger in all 13 (Table 6-1). None of the females of *P. lucida* at the end of any of the trials was as large as the smallest *P. monacha-lucida*. The mean total length of the 35 *P. monacha-lucida* females was 33.9 mm. For the 23 *P. lucida* females it was 26.9 mm, which is about the size at which reproduction begins for both biotypes (Schultz 1961).

The superior growth rate of *P. monacha-lucida* relative to *P. lucida* may be considered heterosis, but first the possibility that its growth rate is only intermediate between parental forms must be ruled out. In a similar experiment (Table 6-2) with *P. monacha*, the unisexuals again grew larger, not quite so conspicuously as when reared with *P. lucida* but, nevertheless, they were larger in four of six trials with a mean of 29.4 to 27.3 (P < .003, Wilcoxon two-sample test).

Table 6-1. Growth rate differences between *Poeciliopsis lucida* (ℓ) and *P. monacha-lucida* (m-ℓ) raised together from birth

Biotype	Trial	Strain	Days	No. at start	Total survivors	Females	Mean length of females (mm)	Mean difference
ℓ	1	VL78-18	96	2	2	1	30.1	5.5
m-ℓ		SV73-7		2	2	2	35.6	
ℓ	2	VL78-18	89	4	4	4	24.1	5.6
m-ℓ		SV73-7		4	4	4	29.7	
ℓ	3	LM78-13	86	5	5	2	25.0	6.6
m-ℓ		SV73-4		5	5	5	31.6	
ℓ	4	M61-31	98	1	1	1	30.8	9.2
m-ℓ		SV73-4		1	1	1	40.0	
ℓ	5	M61-31	94	3	3	1	26.5	9.5
m-ℓ		SV73-4		5	5	5	36.0	
ℓ	6	L74-5	92	2	2	1	26.9	4.5
m-ℓ		T70-3		2	2	2	31.4	
ℓ	7	S68-4	88	4	4	2	28.3	3.0
m-ℓ		T70-3		4	4	4	31.3	
ℓ	8	outcross	89	3	3	3	30.2	4.2
m-ℓ		M61-35		3	3	3	34.4	
ℓ	9	L74-5	92	2	2	2	32.4	7.0
m-ℓ		M65-26		2	2	2	39.4	
ℓ	10	VL78-18	94	2	2	1	26.5	10.2
m-ℓ		M65-24		2	2	2	36.7	
ℓ	11	VL78-18	94	2	2	2	23.7	10.3
m-ℓ		M65-24		2	2	2	33.0	
ℓ	12	VL78-18	96	2	2	2	26.5	8.4
m-ℓ		M65-24		2	2	2	34.9	
ℓ	13	VL78-18	96	2	2	2	24.1	13.9
m-ℓ		M65-24		1	1	1	38.0	

Table 6-2. Growth rate differences between *Poeciliopsis monacha-lucida* (m-ℓ) and *P. monacha* (m) raised together from birth

Biotype	Trial	Strain	Days	No. at start	Total survivors	Females	Mean length of females (mm)	Mean difference
m-ℓ	1	M65-26	98	2	2	2	33.1	2.4+
m		L74-5		2	2	2	30.7	
m-ℓ	2	M61-35	89	2	2	2	30.3	2.8−
m		L74-5		4	3	1	33.1	
m-ℓ	3	T70-3	106	5	2	2	32.5	0
m		outcross		5	5	2	32.5	
m-ℓ	4	SV73-7	93	4	4	4	30.2	3.4+
m		outcross		4	4	4	26.8	
m-ℓ	5	SV73-7	93	5	5	5	26.8	2.0+
m		S68-4		5	5	4	24.8	
m-ℓ	6	SV73-7	95	5	5	5	28.2	3.5+
m		S68-5		5	5	4	24.7	

How the growth rate of a triploid hybrid compares to that of a diploid hybrid is of some interest, not only from the standpoint of interactions in *Poeciliopsis* communities but also as a tool in commercial fish culture, where it has been proposed that polyploidization may produce stocks with improved growth. These growth rate experiments do not offer encouragement to such hopes. When the diploid *P. monacha-lucida* and the triploid *P. monacha*-2 *lucida* were reared together, the diploid grew larger in six out of seven trials (Table 6-3). In spite of having a slower growth rate than a diploid hybrid, the triploid grew faster than *P. lucida* in five out of five trials (Table 6-4).

In all of the above experiments the aquarist's rule of at least "one gallon of water per guppy-sized fish" was adhered to, so that crowding was not a factor. The next series of experiments, which has been extracted from a larger study being done with Valerie Keegan-Rogers, explores the question of growth rate under crowded conditions. The procedure involves rearing recently born young, 1-3 days old, together in 4.5-liter plastic boxes. Members of the two clones or hemiclones are marked by clipping the left or right pectoral fins, after which they are placed together in combinations of 2 X 2 up to 9 X 9. During the first 2 months the young are removed at 10-day intervals, measured under a dissecting microscope, and fin clipped again. Reclipping the fins is necessary while the fish are in their rapid growth because their fins regenerate and clones cannot be distinguished from each other without tissue graft or electrophoretic analysis. With from 4 to 18 fish crowded into a little more than a gallon of water, as they grew larger, a point was reached in each trial at which most or all of the fish of one biotype died. Measurements were compared from the latest possible 10-day interval that would provide a maximum number of survivors.

To illustrate that hemiclones of *P. monacha-lucida* do have different growth rates under crowded conditions, M61-35, which gave evidence of growing more slowly than

Table 6-3. Growth rate differences between *Poeciliopsis monacha-lucida* (m-ℓ) and *P. monacha-2 lucida* (m-2ℓ) raised together from birth

Biotype	Trial	Strain	Days	No. at start	Total survivors	Mean length (mm)	Difference
m-ℓ	1	M65-26	98	5	5	30.9	4.9
m-2ℓ		M61-31		5	4	26.0	
m-ℓ	2	M65-26	89	4	4	28.5	0.4 –
m-2ℓ		M61-31		5	5	28.9	
m-ℓ	3	M65-24	90	1	1	33.1	1.9
m-2ℓ		M61-31		3	3	31.2	
m-ℓ	4	SV73-4	90	5	5	33.4	3.3
m-2ℓ		SV73-4		5	4	30.1	
m-ℓ	5	S68-4	89	4	4	35.6	3.0
m-2ℓ		M61-31		3	3	32.6	
m-ℓ	6	SV73-4	92	6	5	32.6	1.5
m-2ℓ		M61-31		4	4	31.1	
m-ℓ	7	S68-4	90	3	3	35.2	4.2
m-2ℓ		M61-31		3	3	31.0	

others in combination with *P. monacha* (Table 6-2), was tested successively against the two other hemiclones (Table 6-5). It grew more slowly in all four trials with M65-26 and in all five trials with S68-4.

In experiments involving diploid and triploid unisexuals under uncrowded conditions, the diploid grew faster. Under crowded conditions, the two biotypes were rematched, this time choosing two diploids, one the slower growing M61-35 hemiclone and the other the more rapidly growing M65-26 hemiclone; each was matched with a clone of *P. monacha-2 lucida* (Table 6-6). Although the M65-26 diploid grew

Table 6-4. Growth rate differences between *Poeciliopsis lucida* (ℓ) and *P. monacha-2 lucida* (m-2ℓ) raised together from birth

Biotype	Trial	Days	No. at start	Total survivors	Females	Mean length of females (mm)	Difference
m-2ℓ	1	90	3	3	3	35.4	6.4
ℓ			3	3	1	29.0	
m-2ℓ	2	90	5	5	5	32.6	4.6
ℓ			5	5	2	28.0	
m-2ℓ	3	94	2	2	2	38.8	4.9
ℓ			2	2	1	33.9	
m-2ℓ	4	91	2	2	2	35.6	4.7
ℓ			2	2	1	30.9	
m-2ℓ	5	90	5	5	5	32.4	6.4
ℓ			5	5	3	26.0	

Table 6-5. Growth rate differences between three hemiclones of *P. monacha-lucida* under crowded conditions

Trial	Hemiclone	Days	No. at start	Total survivors	Mean length (mm)	Mean difference
1	M65-26	89	7	7	23.7	5.2
	M61-35		7	6	18.5	
2	M65-26	92	7	6	24.5	3.2
	M61-35		7	4	21.3	
3	M65-26	140	4	4	19.0	3.0
	M61-35		4	4	16.0	
4	M65-26	90	2	2	27.0	7.0
	M61-35		2	2	20.0	
1	S68-4	172	9	8	30.0	3.7
	M61-35		9	8	26.3	
2	S68-4	218	6	6	40.5	6.5
	M61-35		6	4	33.0	
3	S68-4	107	2	1	31.0	4.0
	M61-35		2	2	27.0	
4	S68-4	193	5	3	34.3	2.3
	M61-35		5	4	32.0	
5	S68-4	128	2	2	38.5	8.0
	M61-35		2	2	30.5	

larger than the triploid, the M61-35 did not. The apparent slight edge of M61-35 over the triploid is not statistically significant. This is the only experiment discussed in this chapter wherein growth differences between biotypes are not statistically different (Wilcoxon two-sample test); all others are highly significant.

Fecundity

Although fecundity is an important component of fitness, it is one of the most difficult to assess among fishes. Local environmental influences, sometimes in habitats less than half a kilometer apart, may place one population of *Poeciliopsis* at a reproductive peak but leave another completely shut down. Factors such as water temperatures, the amount of sunlight, day length, food supply, and the state of health of the fish, operating alone or in concert, may influence brood size or brood interval or may bring about a total suppression of ovarian activity. In spite of this strong environmental influence, however, reproduction in *Poeciliopsis* is not without some genetic control. Although all 18 or more members of the genus are viviparous, pronounced differences exist among species in terms of ovum size and degree of placental development. These determine the proportion of nutrient support embryos receive directly from the mother vs. from a yolk-laden egg. Egg size, in turn, regulates brood size and

Table 6-6. Growth rate differences between two hemiclones of the diploid unisexual *P. monacha-lucida* and a triploid unisexual *P. monacha-2 lucida* under crowded conditions

Trial	Biotype	Strain	Days	No. at start	Total survivors	Mean length (mm)	Mean difference
1	m-ℓ	M61-35	85	2	2	29.5	3.5
	m-2ℓ	M61-31		2	2	26.0	
2	m-ℓ	M61-35	184	4	3	35.0	1.0
	m-2ℓ	M61-31		4	2	34.0	
3	m-ℓ	M61-35	113	3	3	22.7	0.3–
	m-2ℓ	M61-31		3	3	23.0	
4	m-ℓ	M61-35	200	2	2	34.0	1.0
	m-2ℓ	M61-31		2	2	33.0	
1	m-ℓ	M65-26	244	4	4	41.0	6.0
	m-2ℓ	M61-31		4	3	35.0	
2	m-ℓ	M65-26	128	3	2	34.0	4.0
	m-2ℓ	M61-31		3	2	30.0	
3	m-ℓ	M65-26	153	4	4	36.3	4.0
	m-2ℓ	M61-31		3	2	32.3	
4	m-ℓ	M65-26	166	3	2	41.0	9.0
	m-2ℓ	M61-31		3	2	32.0	
5	m-ℓ	M65-26	190	2	2	36.0	3.0
	m-2ℓ	M61-31		2	2	33.0	
6	m-ℓ	M65-26	190	2	2	30.5	1.5–
	m-2ℓ	M61-31		2	2	32.0	

brood interval (Thibault and Schultz 1978, Turner 1940a, b). Unlike guppies, all members of the genus have superfetation; i.e., they have from two to four different age groups of embryos in the ovary simultaneously; the more stages that exist, the shorter the interval between broods. Reproductively, *P. monacha* is among the least specialized species in that it has only two stages of embryos and no placental arrangement: All nutrients for embryonic development and maintenance are prepackaged in large eggs. *Poeciliopsis lucida* is intermediate in position within the genus in terms of ovarian specialization, having three stages of embryos, eggs of intermediate size, and some maternal support to the developing embryo.

In the hybrids ova sizes reflect parental genome dosages: *P. monacha-2 lucida* has ova slightly larger than *P. lucida*, *P. 2 monacha-lucida* has slightly smaller ones than *P. monacha*, and *P. monacha-lucida* is intermediate between the two parental types (Schultz 1969). Although ovum size clearly reflects a genetic component, the real question is how does it effect fecundity (1) between *P. monacha* and *P. lucida*, (2) among 2n and 3n hybrids, and (3) among their clones and hemiclones, where again the issue of heterosis vs. clonal polymorphism arises?

Because complete brood records are kept on all females in the aquarium room, a massive amount of data are available on fecundity among the various biotypes. Interpreting them correctly, however, has its pitfalls, because separating the environmental influence from the genetic component is no easier when working with laboratory data than it is with field data. Furthermore, fecundity itself is a complex trait, as will be seen in the course of this assessment. In Table 6-7, two criteria have been used that, although related, approach fecundity from different viewpoints. The mean brood size provides only a general picture of the fecundity of a particular biotype under laboratory conditions. Performance varies greatly among individuals, even with identical genotypes, such that successive brood sizes for one may be 2, 9, 16, 16, 9, 12, 24, 12, 2, 2, 1, 2, 1 and for another 2, 3, 2, 2, 4, 2, 2, 3, 1, 2, 1, 3, 1. The first series is typical of a female that has reached her reproductive potential but has fallen off later in her reproductive life either because of age or because she is running out of stored sperm, normally good for about 6 months. Although these fishes mate continuously throughout most of their lives, some females become recalcitrant in middle age. The second series is typical of females which, in spite of having long reproductive lives, never have large broods no matter how well they are treated. Apparently stress imposed during their youth or, perhaps even when they are embryos, can cause permanent damage to their reproductive systems. Consistently low fecundity has been observed in pregnant females brought in from the wild as well as in laboratory stocks. For this reason a second measure has been employed, one that evaluates maximum reproductive potential. Maximum brood size is probably a better estimate of genetic endowment than mean brood size because, although a suboptimal environment can severely suppress phenotypic expression in a trait such as reproduction, there is a limit to how much a superior environment can enhance it. Three values have been examined as supplements to mean brood size: maximum brood size for a biotype; the range of the top 10 broods, each from a different female within a biotype; and the mean of these top 10 broods.

Of three strains of *P. lucida*, two kept in the laboratory since 1961 and one since 1968, one has a lower fecundity than the other two (Table 6-7). Although these lines are maintained by brother-sister matings, they do not suffer from the inbreeding depression characteristic of other inbred organisms. The M61-35 strain of *P. lucida* was always a poor producer and continues to be. The other two, which are comparatively good, are not different from when they were originally started in the laboratory. This same lack of inbreeding depression appears to be true not only of other lines of *P. lucida*, but of other species of *Poeciliopsis* as well. The two strains of *P. monacha* are about equivalent in production. Their general means and the ranges of their top 10 broods are well above those of *P. lucida*. Because of the differences in the reproductive mechanisms between these two species, however, it is important not to reach the hasty conclusion that *P. monacha* is more fecund than *P. lucida*. With three stages of embryos, *P. lucida* has a shorter brood interval (10.3 days) than *P. monacha* (11.6 days), which has only two stages of embryos (Thibault and Schultz 1978). *Poeciliopsis lucida* also reproduces at a slightly smaller size (27 vs. 28 mm) and at a younger age (56 vs. 82 days). In doing so it makes no sacrifice in the length of its reproductive life. In the laboratory *P. monacha* produces only about one-fourth as many young in the winter as it does in the summer. It is considerably more sensitive to the cooler winter temperatures in the aquarium room than *P. lucida*, whose brood production falls only slightly in the winter. The differential response of these two species to temperature are more accurately evaluated in the section, Thermal Stress.

Table 6-7. Comparison of fecundity of *Poeciliopsis lucida* (ℓ), *P. monacha* (m), *P. monacha-lucida* (m-ℓ), and *P. monacha-2 lucida* (m-2ℓ)

Biotype	Strain	Total no. of females	Total young produced	No. of broods	Mean brood size (mm)	Range of 10 largest broods[a]	Mean of 10 largest broods
ℓ	M61-31	74	1820	384	4.7	12-23	13.5
ℓ	M61-35	112	1453	479	3.0	9-14	11.4
ℓ	S68-4	46	1377	246	5.6	10-17	12.2
m	S68-4	65	3010	401	7.5	22-34	26.7
m	S68-5	46	1552	204	7.6	16-30	22.4
m-ℓ	M65-24	27	981	150	6.5	18-38	23.9
m-ℓ	M65-26	30	1059	171	6.2	16-34	22.6
m-ℓ	M61-35	70	1627	384	4.2	12-22	16.1
m-ℓ	Synthesized S68-5	30	419	156	2.7	6-14	9.2
m-ℓ	Synthesized S68-4	30	608	105	5.8	15-23	17.6
m-2ℓ	M61-31	61	1153	344	3.4	8-14	10.4

[a] Each from a different female.

In evaluating fecundity in the unisexual hybrids it is easier to compare them to each other than to the parental species because here, at least, similar reproductive systems are encountered. To illustrate the type of polymorphism that exists in the fecundity of *P. monacha-lucida*, brood data from three hemiclones have been selected (Table 6-7). Two of these hemiclones have about the same mean brood sizes (6.2 and 6.5) and the same mean for the 10 largest broods (22.6 and 23.9); the third, however, has considerably lower values in both categories (mean brood size, 4.2, and mean for the 10 largest broods, 16.1). Interestingly enough, this low score came from the same hemiclone (M61-35, IIIb in Figure 6-1) that performed so poorly in growth experiments. In spite of its poor fecundity record and poor relative growth in the laboratory, it is a well established and widespread hemiclone. First collected at El Cajon in 1961, it was again collected, 12 years later, at San Pedro, a distance by river of 150 km. What adaptations offset its deficiencies has not yet been established. Perhaps in our mate selection experiments, currently in progress, it will prove to be especially capable at acquiring sperm.

Shortly after the S68-4 and S68-5 strains of *P. monacha* were started in the laboratory, two hemiclones of *P. monacha-lucida* were synthesized by crossing them with *P. lucida* (Schultz 1973). Since then, like the wild hemiclones, they have been maintained by backcrossing them with *P. lucida*. One of the hemiclones has had a rather poor reproductive history, with an average brood size of 2.7 and a mean for the 10 largest broods of 9.2; the other has performed more comparably with the wild strains, accruing overall and largest brood means of 5.8 and 17.6 respectively (Table 6-7).

On the basis of fecundity at least, one of these laboratory synthesized hemiclones has a competitive chance of becoming established in nature with the existing wild hemiclones but the second one appears to have a disadvantage that would be difficult to overcome. The reproductive record of a triploid *P. monacha-2 lucida* clone, when compared to the other unisexuals, is also quite low. Its general mean brood size, 3.4, and mean of the largest broods, 10.4, however, are still in the same range as those of M61-35 *P. lucida* and M61-35 *P. monacha-lucida*.

It is perhaps important to be reminded that the significance of these data cannot be extended much beyond establishing that different genotypes among the clones and hemiclones of unisexuals translate into different phenotypes. By looking at collections of wild specimens some insight can be gained, however, as to how these phenotypes will express themselves in nature and what trade-offs in fitness they represent. In one particular series of three collections (Table 6-8), two were made at the same site, 1 year apart, and one from a second site 150 km away. The two clones of *P. 2 monacha-lucida* from these collections were separated from *P. monacha-lucida* and from each other by electrophoresis. However, because in *P. monacha-lucida* electrophoretic phenotypes may include more than one hemiclone and these can be separated only by tissue graft analysis, the diploid unisexuals have been lumped together and attention is directed toward two clones of the triploid *P. monacha-2 lucida*.

At El Cajon, both in 1975 and 1976, clone I was decidedly more successful than clone II by ratios of 18:3 and 12:5, but at San Pedro the order of success was reversed, 6:15. In all three collections, clone II had a larger mean brood size, larger mean total number of embryos, and the greatest number of embryos among the three most fecund females, all of which suggest that clone II is more fecund, but this consistently high fecundity of clone II does not account for the differential success of clones

Table 6-8. Comparison of adult abundance, reproductive potential, and standard length

	Total no. of females	Percentage pregnant	Mean brood size (mm)	Mean total no. embryos	Greatest no. of embryos per female	Mean standard length ± S.D.
El Cajon, 3-26-76						
P. monacha-2 lucida I[a]	18	38.9	3.3	5.1	7, 7, 6	28.6 ± 3.3
P. monacha-2 lucida II	3	100.0	4.6	7.7	12, 7, 4	27.8 ± 4.2
P. monacha-lucida	32	71.9	5.6	8.0	24, 20, 19	28.8 ± 1.9
P. lucida	(8 ♂) 5	100.0	8.4	15.2	36, 19, 10	27.2 ± 6.2
El Cajon, 5-24-75						
P. monacha-2 lucida I	12	75.0	4.5	7.0	17, 15, 9	28.1 ± 4.9
P. monacha-2 lucida II	5	60.0	8.0	13.3	19, 17, 13	33.6 ± 7.4
P. monacha-lucida	14	42.9	15.0	22.5	35, 34, 33	31.5 ± 4.5
P. lucida	(11 ♂) 14	100.0	10.6	19.0	30, 25, 24	29.3 ± 5.0
San Pedro, 5-21-75						
P. monacha-2 lucida I	6	100.0	5.1	10.2	20, 16, 8	33.8 ± 1.8
P. monacha-2 lucida II	15	66.7	9.7	18.4	34, 29, 25	35.0 ± 2.2
P. monacha-lucida	59	94.6	9.5	15.2	31, 30, 29	34.1 ± 2.1
P. lucida	(1 ♂) 25	100.0	10.2	21.6	34, 34, 32	33.5 ± 2.5

[a] *P. monacha-2 lucida* I = M61-31.

between collecting sites. One of the factors to consider here is the differences in the sites themselves. El Cajon is a headwater arroyo with a narrow channel. When torrential rains come, it is subject to scouring and loss of food organisms. The Rio San Pedro, in contrast, has a wide bed 75-100 m across, with a small stream of water flowing through it; it is fully exposed to the sun and enriched by livestock grazing in the river bed. All fish in the Rio San Pedro are in better condition than in the arroyo El Cajon; they are generally larger and have more embryos. This suggests that under favorable conditions, such as provided in the Rio San Pedro, clone II does best; under unfavorable conditions, however, clone I is better adapted.

The three collections are characteristic of numerous others involving the unisexual-bisexual complex in that the unisexual individuals collectively outnumber *P. lucida* and that *P. lucida* has a higher pregnancy rate than unisexuals. This is brought about by the fact that the dominant males of *P. lucida* discriminate against unisexuals. Normally these unisexuals would go unmated except that subordinate males, in their haste to mate before being chased away by dominant males, are inclined to make mistakes (McKay 1971).

Adaptive differences have also been identified between two clones of *P. 2 monacha-lucida* (Vrijenhoek 1978). One clone (clone II) has a higher frequency in downstream areas rich in algae, whereas the other clone (clone I) is better able to invade the depauperate rocky arroyos that are the strongholds of *P. monacha*, its sexual host species. Clone II is more inclined to browse in filamentous algae and detritus; clone I scrapes algae off rocks.

Thermal Stress

Heterosis and clonal polymorphism are especially well illustrated in the response of members of this unisexual-bisexual complex to thermal stress. In a recent study (Bulger and Schultz 1979), the various biotypes were subjected to cold stress, heat stress, and a test for their critical thermal minima. In the acute cold stress experiments, the fish were immersed in water at 1°C for 8 min. They were then returned to the aquarium in which they were acclimated and their survival rates recorded. In the acute heat stress experiments, the fish were exposed to water at 40°C for 30 min and returned to their acclimation tanks. To obtain critical thermal minima the temperature was gradually lowered at a rate of 5°C a day until the fish lost equilibrium.

P. monacha had a lower survivorship than *P. lucida* in both the acute heat and acute cold stress tests but their diploid hybrid derivative, *P. monacha-lucida*, surpassed both parental species in both tests. The test for the critical thermal minima had the same outcome, most *P. monacha* lost equilibrium in the 6-7°C range, *P. lucida* in the 5-7°C range—being somewhat more resistant, and the hybrid in the 4-6°C range, with some individuals still stable at 3°C.

By extending the exposure times to 10 min for the acute cold stress test and 40 min for the acute heat stress test, they could be used to elicit differential responses among hemiclones of *P. monacha-lucida*. The cold stress test indicated that two were cold tolerant, with survivorships of 21/22 and 13/15; two were intermediate, with survivorships 7/14 and 6/14; and two were cold sensitive, both with survivorships of 2/15. In the heat stress test of five hemiclones, three were resistant, with survivorships of 18/21, 10/10, and 10/11, and two were less resistant, with survivorships of 6/16 and 5/11.

The same *P. monacha*-2 *lucida* triploid clone used in the growth rate and fecundity studies was also tested for thermal resistance. In the cold stress test it had a 100% survival rate, the same as the two best clones of *P. monacha-lucida*, but survival to heat stress was worse than for any of the other biotypes (4%).

Conclusions and Summary

When *Poeciliopsis monacha* and *P. lucida* hybridize, instead of producing inviable or sterile progeny as often happens in crosses between species, they produce highly viable and fertile offspring, but all of them are females. Most of these newly launched hemi-clones are probably only temporarily successful at best, but some survive apparently by wedging themselves in between niches not too tightly held by existing biotypes. The selection process that decides which will survive and which will not, does not operate on single genes or on groups of genes and their phenotypes, but on adaptive complexes bound together in the largest possible linkage units, consisting of entire genomes.

These diploid all-female hybrids, *P. monacha-lucida*, reproduce by hybridogenesis, transmitting only monacha chromosomes to haploid eggs and acquiring a new lucida genome for each new generation. Two triploid biotypes, *P. monacha*-2 *lucida* and *P.* 2 *monacha-lucida*, have arisen, apparently from the diploids. Both reproduce clonally by gynogenesis from triploid eggs.

Three life history characters (growth rate, fecundity, and resistance to thermal stress) have been chosen for consideration in this study because they are believed to be especially important in the relationships between the parental species and the unisex-uals and among the clones and hemiclones of the unisexuals.

Resistance to thermal stress and growth rate under both crowded and uncrowded conditions reveal that the diploid unisexual hybrids are superior to the parental species and that substantial adaptive differences exist among hemiclones. In fecundity, also, most hemiclones are superior to *P. lucida* with which they live, and when to this is added the capacity to produce only female progeny, a considerable overall reproduc-tive advantage exists. Whether or not these hemiclones are reproductively heterotic relative to their other parental species, *P. monacha*, is clouded by the fact that we are comparing two quite different reproductive systems, one (*P. monacha*) that includes fully yolked eggs vs. one that depends on a placental arrangement to nourish embryos.

One triploid clone, when compared to the diploid unisexuals, did not reveal that the addition of an extra genome provided any special properties. In general it was less fecund, had a slower growth rate, and was less resistant to thermal stress than diploid unisexuals. It compared well to the parental species in growth rate experiments, how-ever, surpassing both *P. monacha* and *P. lucida*. Its survivorship to acute cold stress was 100%, ranking it with the top two hemiclones of the diploid, but its survival rate under heat stress was lower than any of the biotypes tested.

The success of the various biotypes of this bisexual complex varies markedly from one locality to the next and from one season to the next (Angus and Schultz 1979, Schultz 1977, Thibault 1978). How successful each clone becomes is likely to depend to a large extent on how much of its specific niche is available. Unfortunately, such measurements in nature are not easily obtained. Ultimately it may be found that the overall abundance of a unisexual species depends on how many clones it comprises and how much combined niche they can capture and hold.

Acknowledgments. I thank Judy Sullivan, our aquarist, for her never-ending vigil over the fish colony, for her close participation in the experiments, and for reading the manuscript. I also thank Valerie Keegan-Rogers for allowing me to use some of our data from a larger, more specific project that is to be published jointly upon its completion, and Ellie DeCarli, for typing the manuscript. I am grateful to Robert Vrijenhoek and James Leslie, who allowed me to dissect specimens they collected and identified electrophoretically (Table 6-8). This work is supported by the National Science Foundation, DEB-7725341.

References

Angus, R. A., Schultz, R. J.: Clonal diversity in the unisexual fish *Poeciliopsis monacha-lucida*: a tissue graft analysis. Evolution 33, 27-40 (1979).

Bulger, A. J., Schultz, R. J.: Heterosis and interclonal variation in thermal tolerance in unisexual fishes. Evolution 33, 848-859 (1979).

Cimino, M. C.: Meiosis in triploid all-female fish *Poeciliopsis* (Poeciliidae). Science 175, 1484-1486 (1972).

McKay, F. E.: Behavioral aspects of population dynamics in unisexual-bisexual *Poeciliopsis* (Pisces: Poeciliidae). Ecology 52, 778-790 (1971).

Moore, W. S.: Components of fitness in the unisexual fish *Poeciliopsis monacha-occidentalis*. Evolution 30, 564-578 (1976).

Moore, W. S.: A histocompatability analysis of inheritance in the unisexual fish *Poeciliopsis* 2 *monacha-lucida*. Copeia 1977, 213-223 (1977).

Schultz, R. J.: Reproductive mechanisms of unisexual and bisexual strains of the viviparous fish *Poeciliopsis*. Evolution 15, 302-325 (1961).

Schultz, R. J.: Hybridization experiments with an all-female fish of the genus *Poeciliopsis*. Biol. Bull. 130, 415-429 (1966).

Schultz, R. J.: Gynogenesis and triploidy in the viviparous fish *Poeciliopsis*. Science 157, 1564-1567 (1967).

Schultz, R. J.: Hybridization, unisexuality and polyploidy in the teleost *Poeciliopsis* (Poeciliidae) and other vertebrates. Am. Nat. 103, 613-619 (1969).

Schultz, R. J.: Special adaptive problems associated with unisexual fishes. Am. Zool. 11, 351-360 (1971).

Schultz, R. J.: Unisexual fish: Laboratory synthesis of a "species." Science 179, 180-181 (1973).

Schultz, R. J.: Evolution and ecology of unisexual fishes. In: Evolutionary Biology. Hecht, M. K., Steere, W. C., Wallace, B. (eds.). New York: Plenum Press, 1977, pp. 277-331.

Thibault, R. E.: Ecological and evolutionary relationships among diploid and triploid unisexual fishes associated with the bisexual species, *Poeciliopsis lucida* (Cyprinidontiformes: Poeciliidae). Evolution 32, 613-623 (1978).

Thibault, R. E., Schultz, R. J.: Reproductive adaptations among viviparous fishes (Cyprinodontiformes: Poeciliidae). Evolution 32, 320-333 (1978).

Turner, C. L.: Pseudoamnion, pseudochorion and follicular pseudoplacenta in poeciliid fishes. J. Morphol. 67, 58-89 (1940 a).

Turner, C. L.: Superfetation in viviparous cyprinodont fishes. Copeia 1940, 88-91 (1940 b).

Vrijenhoek, R. C.: Genetic relationships of unisexual-hybrid fishes to their progenitors using lactate dehydrogenase isozymes as gene markers (*Poeciliopsis*, Poeciliidae). Am. Nat. 106, 754-766 (1972).

Vrijenhoek, R. C.: Coexistence of clones in a heterogeneous environment. Science 199, 549-552 (1978).

Vrijenhoek, R. C., Angus, R. A., Schultz, R. J.: Variation and clonal structure in a unisexual fish. Am. Nat. 112, 41-55 (1978).

Chapter 7

The Evolution of Life Span

CHARLES E. KING*

Introduction

In 1954, Lamont Cole stated the basic tenet of life history pattern analysis and in so doing established a new branch of evolutionary ecology. According to Cole, "The total life history pattern of a species has meaning in terms of its ability to survive and ecologists should attempt to interpret these meanings." The interpretational method he proposed and used was "to compute the characteristics of the future hypothetical population by assuming an unvarying pattern of the life history features which govern natality and mortality."

Although Cole's study attracted considerable attention, the interpretation of life history patterns languished until the publication of R. C. Lewontin's paper in *The Genetics of Colonizing Species* (1965). Lewontin's contribution was to introduce classical population dynamics to the analysis. He also brought the problem to the attention of individuals in the nascent field of population biology who were schooled in the interpretation of ecological patterns in an evolutionary context. These developments have stimulated, either directly or indirectly, most of the dominant conceptual movements in ecology over the past 15 years, including r and K selection, age-specific selection, selection for competitive ability, and most recently the analyses associated with life history "strategies" and "tactics."

When considering a constant environment, the usual procedure has been to define two or more alternative life history patterns. These patterns are then examined for their effects on fitness, commonly r, the instantaneous rate of population growth. With simple patterns, the assumption is made that the life history maximizing r will be the superior adaptation and will therefore be favored by natural selection. Analyses of this type have led to the conclusion that selection should minimize the age of first reproduction, and maximize both fecundity rates and survival to reproductive maturity. A corollary suggestion is that age-specific reproductive value is maximized by the superior life history pattern (Schaffer 1974, Taylor *et al.* 1974), but a recent analysis by Caswell (1980) casts doubt on the generality of this assertion.

*Department of Zoology, Oregon State University, Corvallis, Oregon 97331 U.S.A.

In other studies the focus has been on optimization of the components of fitness. For instance, Lack (1947) proposed that clutch size in certain birds is adjusted by natural selection to maximize reproductive success consistent with the expected availability of food for nestlings and the ability of the parents to exploit the food resources. In birds or other organisms in which the clutch size is small, repeated (iteroparous) reproduction in successive years is an important component of fitness. Clutch size maximization in one year may produce physiological stress and lower parental survival in future years. Consequently, fitness may be highest when clutch sizes are optimized rather than maximized. This topic has been considered recently by Stearns (1976) and De Steven (1980).

Another example of life history optimization is derived from the work of my colleague, Peter S. Dawson. In laboratory populations of the flour beetle *Tribolium castaneum*, oviposition and pupation are frequently synchronized by culture techniques that impose discrete generations on the populations. Larval flour beetles are cannibalistic on pupae. In a dense, synchronized population of developing larvae, the first individuals to pupate have substantially higher probabilities of being cannibalized than individuals pupating later, when there are fewer larvae (and more pupae) in the experimental container. Thus, in spite of the general conclusion that fast development maximizes fitness, in this situation the timing of pupation appears to be determined by a stabilizing selection that results from countering forces favoring fast and slow developmental times (Dawson 1975).

Although there are many hypotheses on the causes of aging and senescence (see, for instance, reviews by Lamb 1977, Lints 1978), with the exception of an argument rooted in group selection by Wynne-Edwards (1962), none of these suggests that life span is a primary life history feature the duration of which is subject to the direct action of natural selection. The hypotheses on aging treat death as either a secondary (indirect) consequence of selection for optimal reproductive patterns, or as an environmental rather than a genetic event. In the former situation, postreproductive survival is not explained and in the latter situation, unless there is some genetic control over the effects of accidents, life span is simply a statistical happenstance. In this chapter I will argue that these views are too restrictive. I suggest that life span is, in fact, directly related to fitness and, moreover, that it can be subject to direct selection rather than just indirect selection operating through "trade-offs" between reproduction and survival during the reproductive period as discussed by Hamilton (1966) and many others.

Materials and Methods

In response to appropriate environmental cues, monogonont rotifers have normal meiosis and sexual recombination. In the absence of these cues, reproduction occurs by diploid, ameiotic parthenogenesis and unless mutation occurs, the parthenogenetic descendents of a single female constitute a genetically homogeneous clone. Details of this life cycle are presented in Gilbert (1977) and King (1977, 1980). Reviews of studies of aging using rotifers are available in King (1969) and King and Miracle (1980).

Experimental data on which this chapter is based were obtained using parthenogenetic females of clone SP (King and Miracle 1980) of the marine rotifer *Brachionus plicatilis*. Because both survival and fecundity rates are strongly influenced by environmental parameters, experimental conditions are presented in some detail below.

A number of studies have shown that rotifer life spans are influenced by parental age (King 1967). For that reason it is important to use orthoclones in which parental age is held constant in successive generations by discarding all progeny produced by individuals not of the desired age. In the present study, replicated life table analyses were conducted by isolating each of 36-72 third-generation neonates of a 3-day parental-age orthoclone in multiwell tissue culture plates (Falcon 3008) containing a 0.5-ml food suspension of either B or Z culture medium. Medium B contained 300 mM NaCl, 94 mM NaHCO$_3$, 9.7 mM H$_3$BO$_3$, 10 mM KCl, 10 mM MgSO$_4$ · 7H$_2$O, 4.7 mM MgCl$_2$ · 6H$_2$O, 1.2 mM NaNO$_3$, 0.04 mM NaH$_2$PO$_4$, 1 ml of trace metal solution TMS-I, and 1 ml of vitamin mix. The pH of medium B was adjusted to 8.0 using either HCl or NaOH. Medium Z contained 400 mM NaCl, 2 mM NaHCO$_3$, 10 mM KCl, 10 mM MgSO$_4$ · 7H$_2$O, 9.4 mM MgCl$_2$, 1 mM NaNO$_3$, 0.04 mM NaH$_2$PO$_4$, 10 mM CaCl$_2$, 2 mM tris, 5 mM glycylglycin, 1 m. TMS-I, and 1 ml vitamin mix. Medium Z was adjusted to pH 7.8. TMS-I contained 35 μM ZnSO$_4$ · 7H$_2$O, 10 μM MnCl$_2$ · 4H$_2$O, 5 μM Na$_2$MoO$_4$ · 2H$_2$O, 0.3 μM CoCl$_2$ · 6H$_2$O, 0.3 μM CuSO$_4$ · 5H$_2$O, 2 μM Fe citrate, 48 μM EDTA (Na$_2$ salt), and 400 μM H$_3$BO$_3$. The pH of TMS-I was adjusted to 7.0. One liter of vitamin mix contained 200 mg thiamin HCl, 1.0 mg biotin, and 1.0 mg vitamin B$_{12}$.

All experiments were conducted at 25°C in a constant temperature room with a 16L:8D cycle. Algae (*Dunaliella tertiolecta*, University of Texas Culture Collection No. 999) were cultured under the same conditions and were enumerated by Coulter counter after cells had been removed from 3-day-old cultures by centrifugation and resuspended in fresh, Millipore-filtered, culture medium. Experimental animals were transferred to fresh algal suspensions once each day, at which time each was examined for viability and for number of progeny produced during the preceeding 24-hr period. Only amictic (parthenogenetic) females were used to calculate population dynamics.

Two measures of the duration of life can be derived from these experiments. The first is simply the average life span (LS) of all individuals in the cohort. The second is the L$_{50}$ or 0.5 l$_x$ measure indicating the age at which 50% of the cohort is alive and the other 50% is dead. This second measure is designated mean survivorship. Because all of the survivorship curves reported in this chapter are approximately rectangular, the death phase for each cohort is approximately linear and the two measures are quite similar (0.5 l$_x$ = –0.364 + 1.052 LS with a correlation of 0.99).

Population Dynamics

A total of 1478 individuals divided into 36 treatment groups was examined at daily intervals to determine the population dynamics reported in this chapter. Culture medium was varied in some experiments and food level was varied in others. The influence of culture medium was initially examined at a constant food level of 500,000 cells per milliliter. The results presented in Figure 7-1 were obtained for the four combinations of rearing algae and rotifers on media B and Z. Although substantial effects of culture medium are apparent, the focus of this chapter is on the relative variability of the response to the four environments. The range in instantaneous population growth rate, r, from these life table experiments is 0.62-1.14 and has a mean ± S.E. of 0.92 ± 0.038

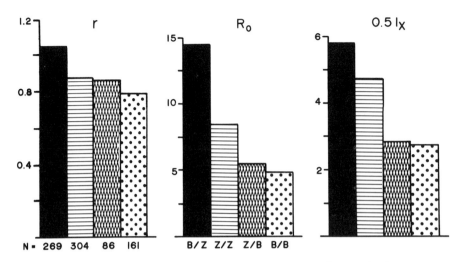

Figure 7-1. The influence of algal and rotifer culture media on the instantaneous growth rate (r), net reproduction (R_0), and mean survivorship (0.5 l_x) of the rotifer *Brachionus plicatilis*. Two media were used, B and Z (see text); the ratio presented under the center histogram is of algal medium to rotifer medium. Numbers of individuals in the four life table experiments are indicated under the left histogram.

and a coefficient of variation of 17.7%. For net reproduction, R_0, the range is 3.8-19.9 offspring per female (9.6 ± 1.19) and the coefficient of variation is 52.3%. Finally, mean survivorship, 0.5 l_x, varies from 2.6 to 6.9 days (4.5 ± 0.33) and has a coefficient of variation of 31.3%. If fitness is defined as the rate of contribution of offspring to the next generation, it is clear that growth rate and not net reproduction of life span is the appropriate fitness measure. Figure 7-1 demonstrates that rather similar growth rates can be obtained from quite dissimilar net reproductive rates and life spans.

This point is demonstrated even more clearly in the results presented in Figure 7-2. The upper panel depicts patterns of survivorship (l_x) obtained in two experiments performed using different food levels and rotifer culture media. The two experiments differed in both food level and medium (500,000 cells/ml and medium B for the low survivorship treatment vs. 62,500 cells/ml and medium Z for the high survivorship treatment). Each of the survivorship curves is negatively skewed and thus indicates a "physiological" life span (King and Miracle 1980). Both experiments produced identical instantaneous population growth rates (r) of 0.84. It should be noted, however, that mean survivorship differs by a factor of 3.5 in the two experiments. The age-specific fecundity (m_x) patterns obtained in these experiments are presented in the bottom panel of Figure 7-2. Net reproduction ($\sum_{x=0} l_x m_x$) differs by a factor of approximately 3 in the two environments.

This substantial variation in age-specific mortality and fecundity is remarkable when it is recalled that both experiments were performed with the same clone. Moreover, the change in survivorship and fecundity patterns is not accompanied by a loss of fitness because both cohorts had identical growth rates. What is being shown here, then, is response variability of a single genotype and not genetically differentiated populations.

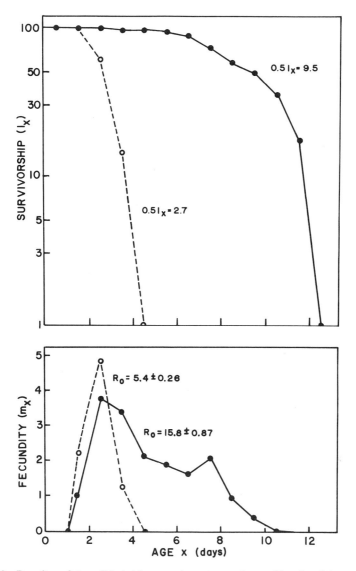

Figure 7-2. Results of two life table experiments on the rotifer *Brachionus plicatilis* performed with different media and quantities of food. Associated values for mean survivorship in days and net reproduction (R_0 = average number of offspring per female lifetime) are presented on the graph.

The concept of reproductive value has been used in several contexts for study of senescence and life history evolution. Hamilton (1966), and more recently Charlesworth (1980), have demonstrated that reproductive value is not an appropriate index of the intensity of selection on genes having age-dependent effects. Caswell (1980) has concluded from a theoretical analysis that maximizing mean fitness is not necessarily equivalent to maximizing age-specific reproductive value. Figure 7-3 presents

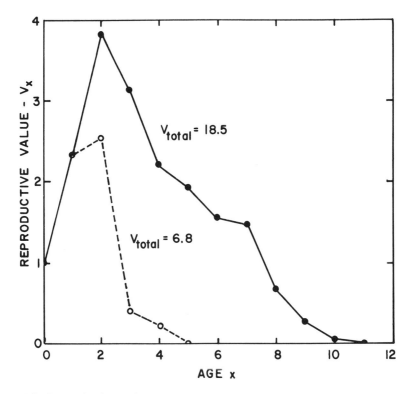

Figure 7-3. Reproductive value curves for the experiments presented in Figure 7-2. Age is measured in days. V_{total} = area under the curve.

reproductive value curves for the two life table experiments under consideration. Both curves have maxima at 2.0 days. In spite of the identity of growth rates, the two curves differ in age-specific reproductive values for different ages, and in their total areas (V_{total}). These observations cast further doubt on the significance of reproductive value as a measure of fitness, although they do not affect utilization of reproductive value measures as a response variable to environmental change. Again, however, I emphasize that the two curves depict phenotypic responses to environmental variation and not genotypic differences, because the same clone was used in both experiments.

Figure 7-4 provides the information that explains how two sets of survivorship and fecundity measures as different as those in Figure 7-2 can produce the same growth rate. Each age class of a life table cohort in which the age-specific fecundity, m_x, is positive makes a contribution to the cohort's growth rate. The ultimate growth rate, r, is the sum of all age-specific contributions to r. It is therefore possible to study the pattern of acquisition of r by progressively summing the proportionate age-specific contributions from age 0. The striking differences in both survivorship and fecundity that were presented in Figure 7-2 are not reflected in the rate of acquisition of the ultimate growth rate (r = 0.84 in each). Although individuals in the cohort depicted by the dashed line have short life spans, they also have higher rates of early reproduction than individuals in the cohort depicted by the solid line. Reproduction was initiated in

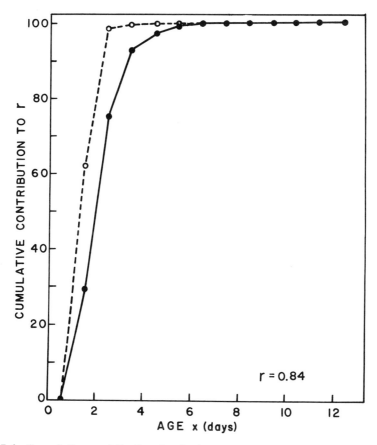

Figure 7-4. Cumulative contribution to the instantaneous population growth rate (r) of females in each age class of the experiments presented in Figure 7-2.

both cohorts at an age of 1.5 days. Approximately 98% of the total value of r is attained by the short-lived cohort in the first 2.5 days of life. During this period, the average neonate produces 5.2 offspring; in the same period a neonate in the long-lived cohort produces 4.8 offspring. Thus, the growth rates of the two cohorts are the same in spite of the great difference in their net reproductive rates because late reproduction in the long-lived cohort compensates for the relatively small deficiency in its early reproduction. Such effects are well known, but they illustrate a point that deserves more emphasis. This point is that when population growth rate is high, reproduction late in life may have no selective significance.

The period required to reach 95% of the ultimate growth rate of r = 0.84 is 2.4 days for the short-lived cohort and 4.0 days for the long-lived cohort (Figure 7-4). In each of the two cohorts, reproduction influencing fitness is completed early in life. Late reproduction, therefore, does not meaningfully affect population growth rate. It is not a component of fitness as measured by r.

To broaden the data base, results from 36 separate life tables are presented in Figure 7-5. These experiments were conducted with a variety of different media (calci-

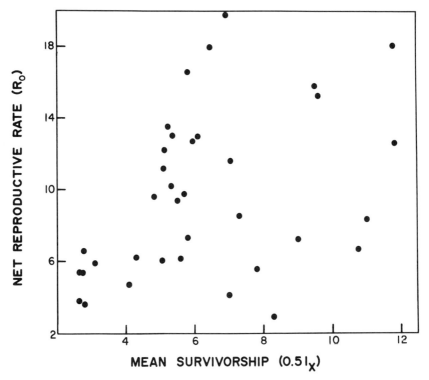

Figure 7-5. The relationship between mean survivorship and net reproductive rate in a series of 36 separate life table experiments with *Brachionus plicatilis*. Experiments differed from each other in culture medium of both rotifers and algae, and in quantity of algal food.

um supplemented and unsupplemented B and Z) under a variety of different food concentrations (ranging from 3900 to 1,000,000 cells/ml). All were performed using the same clone, SP, at 25°C. The correlation between mean survivorship and net reproduction in these experiments is 0.318 and is not significant by t test. The coefficient of determination indicates that only 10% of the variation in net reproduction is explained by variation in mean survivorship. Although positive relationships between net reproduction and survivorship have been found in studies employing different conditions and rotifer species (Snell and King 1977), the data for *B. plicatilis* suggest that life span and number of offspring produced are independent. Population growth rate is determined by survival to reproductive age and by early reproduction.

Is Life Span an Adaptive Character?

Two important conclusions may be drawn from the experiments just presented:

(1) No clear relationship was detected in this diverse series of environments between mean survivorship and net reproductive rate. Long-lived individuals tended to

have large numbers of offspring if they were well fed, and small numbers if they were not.

(2) Late reproduction has little influence on population growth rate.

Taken together, these conclusions may lead to the suggestion that life span is not an adaptive character when growth rate is high. I will now argue that this notion is incorrect and that life span is a direct consequence of stabilizing selection operating during the period of senescence. This argument will lead to the hypothesis that in many organisms with iteroparous reproduction, life span is expected to equal twice the generation time.

Four life history patterns are presented in Figure 7-6. Organisms such as bacteria and ciliates that reproduce by fission, or ones such as pacific salmon and mayflies that reproduce with a "big bang" pattern, have discrete generations and die immediately after reproduction. Generation time (T) in such circumstances can be defined as the temporal interval between identical life history points in two successive generations. In both of these situations life span is precisely equal to generation time and is therefore a direct determinant of population growth rate and fitness.

The bottom panels of Figure 7-6 present two life history patterns in which life span is not equivalent to generation time as defined in the preceding paragraph. If a period of diapause occurs between parental reproduction and environmental exploitation by the offspring, the elapsed time between birth and death of the parent may be considerably shorter than the generation time. This is true whether or not all reproduction occurs at a single point (as shown) as long as the period of diapause is longer than the period between birth and death. Excluding the period of embryonic development, the life span of mictic (sexually reproducing) rotifers is generally about the same as it is for those reproducing by parthenogenesis (King 1970). Sexual eggs have hatching times that may be orders of magnitude longer than asexual eggs, and frequently during winter or other periods of adverse environmental conditions these eggs are the only surviving life history stage for the entire population. Again, however, there is no difficulty in understanding this relationship between life span and generation time because the ecological and physiological significance of the period of diapause is usually obvious.

The concept of generation time is more abstruse for organisms having multiple episodes of reproduction and overlapping generations (Leslie 1966). In these organisms, generation time can be defined as the period required for a population growing at rate r to increase by the factor R_0. In quantitative terms, $T = \ln R_0/r$. This expression presumes a mathematical concentration of all reproduction at point T in the life table of a cohort reproducing at rate r.

Iteroparous reproduction is illustrated in the lower right panel of Figure 7-6. This pattern is found in amictic rotifers (including those in the present study) and most vertebrates (including people). It produces both overlapping generations and life spans that are longer than generation times. If the above definition of generation time is accepted all reproduction is considered to be concentrated at age T; it is obvious that an individual's primary and direct contribution to fitness will be made in the period from its birth to T. However, does this type of mathematical manipulation make biological sense? The primary characteristic of iteroparous reproduction is that all offspring are not produced at the same time. Using an analytical tool that concentrates all

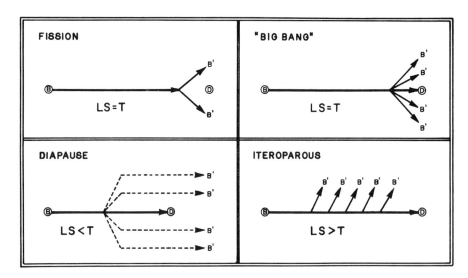

Figure 7-6. Relationship between life span (LS) and generation time (T = 1n R_0/r) in four different life history patterns. B, birth time of parent; B′, birth times of offspring; D, death time of parent.

reproduction at T is effectively assuming that the spread or variance in timing of reproduction is irrelevant except in so far as this variance affects the magnitude of T. If fitness is determined by reproduction occurring between birth and age T, however, then reproduction subsequent to T can be ignored. The validity of this process can be examined by measuring the proportion of the total rate of increase, r, that is realized by age T. This calculation is presented in Figure 7-7 for the 36 life table experiments presented earlier. In spite of the fact that T is always less than the age of median offspring production in a growing population, it can be demonstrated that approximately 85% of the total r is determined by reproduction before T. This approximation is true over a wide range of r, R_0, T, and 0.5 1_x values. Therefore, although the mathematical concentration of all reproduction at T is highly artificial, reproduction after age T contributes relatively little to fitness.

Senescence will be defined for my purposes in this chapter as starting at age T and continuing to death. During senescence in a growing population, there is little additional direct contribution to r because most effective reproduction has already been completed. However, any form of behavior operating during senescence that increases the probability of grandchildren will be favored by selection at the individual level. To explain the evolution of senescence and life spans that substantially exceed T, some form of behavior must be identified that increases the probability of grandchildren. The behavior that meets this criterion is parental care. Life span should increase in response to this selection until a countering force acts to reduce the probability of grandchildren.

The major feature of parental care behavior is its beneficial influence on survival and reproduction of the young. The evolutionary reward of such behavior to the parent is an increased number of grandchildren. However, parental care extended beyond birth of the grandchildren can produce competitive interference, as discussed

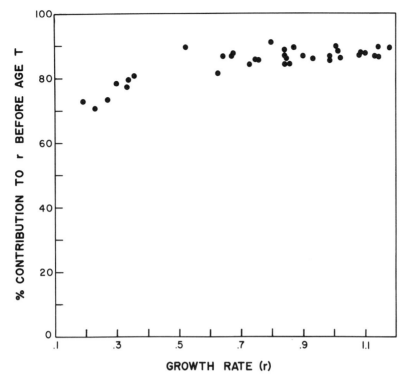

Figure 7-7. Proportionate contribution to total instantaneous growth rate (r) made during the period between birth and age T as a function of r in the 36 life table experiments presented in Figure 7-5.

below, and result in a reduced number of grandchildren. Therefore, the duration of the parental care period is expected to be optimized by stabilizing selection. Moreover, its limits should be approximately coterminous with the period of senescence.

My hypothesis on the evolution of life span in iteroparous organisms can be stated as follows: given that r is positive, most offspring significantly influencing fitness are born to a parent by age T in the parent's life. For these offspring to develop, mature, and reproduce requires another period of duration T. Parental care, therefore, should also occupy a period of length T. Under this hypothesis, life span is divided into two periods, each of which has duration T. Parental death is expected to occur at age 2T when the parent's offspring have completed most of their significant reproduction. The point "2T" is determined by stabilizing selection.

I consider parental care to include any form of parental behavior that increases the survivorship or fecundity of the offspring. Two types of parental care may be recognized, direct and indirect. Direct parental care, such as the feeding or the protecting of children, does not affect the probability of offspring production because the offspring already have been born. It does, however, affect the survival of offspring and serves to increase the probability of grandchildren.

Neither rotifers nor many other organisms show obvious parental care behavior. There are, however, numerous forms of indirect parental care that are mediated through environmental conditioning. (Some of these behaviors substantially distort the usually conotation of the term "parental care." However, if they meet the criterion of increasing parental fitness by increasing the probability of grandchildren, for want of a more appropriate term they are referred to here as indirect parental care.) In *Brachionus calyciflorus*, Gilbert (1963) experimentally demonstrated that the cue for switching from asexual to sexual reproduction is a substance released to the medium by the rotifers themselves. Production of the cue is by amictic females reproducing asexually; the effect of this production is displayed in the next generation when the mictic daughters reproduce sexually. King and Snell (1980) have demonstrated density-dependent induction of sexual reproduction in field populations of *Asplanchna girodi*. Results of this study are consistent with Gilbert's (1963) mechanism, although the field study did not include a bioassay for a chemical inducer. However, there is evidence (King 1972, 1977) that not all cooccurring clones respond at the same time to a cue inducing sexual reproduction. The possibility exists, although it has not been experimentally demonstrated, that different clones produce or respond to chemically different inducers. If such specificity can be demonstrated, then the mixis response not only insures perpetuation of the parent's genes, it meets the criterion of a behavior that is targeted at genetic replicas of the parent and that increases the parent's inclusive fitness.

In this example, a parent rotifer reproduces and modifies the environment in which its progeny develop. As a result of that modification, the ensuing generation produces sexual resting eggs that are able to withstand harsh and inhospitable environments. The fitness of the parent is thus substantially enhanced by its indirect parental care behavior of releasing a mixis cue to the environment.

Many studies have demonstrated that offspring may have enhanced survival when they remain associated with their parents. Such a demonstration would be particularly interesting in the context of the current chapter if it involved a species reproducing by apomictic parthenogenesis. A possible example is given by Hamilton (1966) in his discussion of the association of postreproductive aphids with their clonal offspring. The presumed benefit of this association may be mediated through the food supply.

Numerous examples of both chemical and physical conditioning of the environment by organisms can be found in most major ecology textbooks. If parental activity either maintains or improves the environment inhabited by offspring, and if the environmental amelioration maintains or improves the reproductive potential of these offspring, then such parental activity can be regarded as indirect parental care. Few studies have focused on this aspect of parent-offspring associations; however there are numerous examples of site fidelity (philopatry) extending across generations in birds and mammals (see Greenwood 1980 for a recent review). Individuals of one sex in the offspring group may remain associated with the parents, while the other sex disperses. In the Florida scrubjay (Woolfenden 1975), it is the male that remains on the parental territory and some groups contain males of several different year classes. The reward to both parents and offspring of this association can apparently take two forms. First, males help rear younger siblings and thus receive a selective benefit even if they do not breed. Second, male offspring may replace their fathers if death occurs, or they may establish their own territories from a portion of the parental territory. Direct parental care obviously occurs in birds prior to fledging of the young. In the Florida scrubjay example, permitting male offspring to remain on the parental territory after they have reached breeding age can be viewed as an indirect form of parental care. To the extent that such indirect care by senescing parents ultimately enhances the reproduction of offspring, increased life span should be favored by natural selection.

Counterbalancing these trends to increase parental lifespan is the deleterious effect parents can have on their offspring once reproductive maturity has been attained. Intraspecific competition is generally more intense than interspecific competition because of the high genetic and therefore morphological congruity among the competitors (King 1972, Snell 1979). There exists, therefore, a strong potential for competition between parents and their offspring. Young individuals tend to be rather poor competitors, but in many species parents and their offspring exploit different resources. This diversification is particularly notable in the diets of most aquatic animals. However, as young individuals approach maturity their diets become more like that of their parents and the potential for competition increases. Active competition between parents and their maturing offspring is expected to reduce the probability of grandchildren. In the absence of dispersal, one means to avoid this competition is for parents gradually to withdraw from the competitive environment as their offspring mature. Such a solution is provided by senescence and the progressive loss of the parent's ability to exploit the shared environment. A second solution, of course, is death of the parent when continued life means lower fitness. That point, I suggest, is approximately twice the generation time.

I return now to the set of 36 life tables introduced earlier. These data were obtained in a broad spectrum of different chemical environments under a variety of nutritional regimes. Mean survivorship varied from a low of 2.7 days to a high of 11.8 days in these experiments. According to my hypothesis the relationship between mean survivorship and generation time for this data set should produce a slope of 2. As shown in Figure 7-8, the observed slope is 1.99. In addition, the correlation between these variables is 0.79 and its difference from zero highly significant (p < 0.001) by t-test.

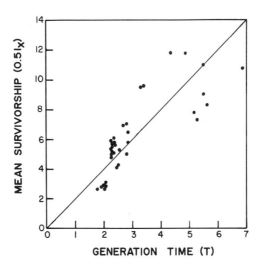

Figure 7-8. Relationship between mean survivorship (0.5 l_x) and generation time (T) in the 36 life table experiments presented in Figure 7-5.

Discussion

Although the agreement between observed and expected life spans for this data set is remarkable, a substantial amount of skepticism concerning the validity of the hypothesis should be exercised until a broader series of tests involving different taxonomic groups has been undertaken.

There are several conditions that must be met before the proposed hypothesis can be applied to a data set. Foremost of these is that the population being examined must have a negative skew rectangular survivorship curve. Departures from a negative skew in a homogeneous population indicate the effects of random, or nonphysiological mortality sources. Diagonal survivorship curves do not imply that senescence is absent in the population. They do indicate substantial mortality from random events of accidents that are not genetically programmed. Effects on mean life span from this category of mortality sources are outside the domain of my hypothesis.

In addition, parental care effects are not expected to occur if there is dispersal and parents and their offspring occupy different habitats (such as in "fugitive" or colonizing species). Another caution relates to the construction of life tables based only on females and their female offspring. Variation in sex ratios can influence the calculated rates, and males and females may differentially contribute to parental care. Neither of these constraints influences the present data set but either may influence others.

Another concern relates to a statistical problem. Although the presented hypothesis states that life span is expected to be twice the generation time, the physiological life span of an individual is subject to many influences that cannot be predicted or understood precisely. Mean life spans, even in a population having a negative skew survivorship curve, always have associated variances. These variances may lead to rejection of the hypothesis when it should be accepted, or acceptance when it should be rejected.

A second statistical concern is that the observed slope of 1.99 presented in Figure 7-8 may be an artifact created by averaging data from populations with low and high growth rates. It has already been shown in Figure 7-7 that even in populations having low instantaneous growth rates, most of r is determined by age T. A more direct way to examine this potential criticism is to ask whether there is a significant relationship between the ratio of observed/expected (= 2T) mean survivorship and the instantaneous growth rate, r. This information is presented in Figure 7-9. The coefficient of determination between the illustrated variables is 5.2% and the slope of a regression of the growth rate on the ratio does not significantly differ from zero. It is therefore reasonable to conclude that the relationship is not an artifact of averaging.

It is important to recognize that this hypothesis does not constitute a new theory of senescence. Although theories not based on some form of genetically "programmed" senescence are obviously excluded, the proposed hypothesis addresses the ultimate evolutionary timing of life span rather than its proximate physiological determination. In this sense, my hypothesis is similar to those of Williams (1957) and Medawar (1957) on the evolution of senescence. These individuals suggested that senescence was a result of incorporation of pleiotropic genes having different effects on reproduction at different ages. Charlesworth (1980) has reviewed these hypotheses recently and Rose and Charlesworth (1980) have published experimental support for the role of pleiotropic genes in senescence.

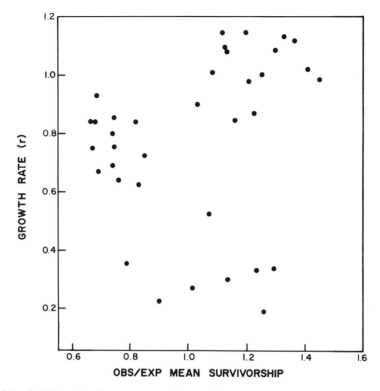

Figure 7-9. Relationship between growth rate (r) and the ratio of observed to expected mean survivorship in the 36 life table experiments presented in Figure 7-5. Expected mean survivorship is twice the generation time (T).

Hamilton (1966), commenting on Williams's paper, suggested that new genes having net beneficial age-specific effects (positive early-negative late in the reproductive period) would cause an increase in population growth rate. The increased growth rate would be countered by density-dependent factors and reacquisition of fecundity and mortality patterns producing a stationary population. My hypothesis is based on a similar scenario, but in contrast to that of Williams, it relates to parental behavior during the period of senescence. Consider a stationary population in which there is no parental care. A new mutant that elicits parental care subsequent to reproduction is expected to increase the number of grandchildren (but not the number of children). The rate of increase for individuals bearing the new gene is positive with respect to other genotypes lacking parental care. This, in turn, will produce selection to increase life span and extend the period of parental care. Pleiotropy is also involved with the postulated mutant because if parental life is extended beyond birth of the grandchildren, competition is expected to reduce the net beneficial effect of the new gene. Both Williams's and my hypotheses involve pleiotropic effects and stabilizing selection. Williams's proposal, however, relates to the onset of senescence, whereas my proposal relates to the termination of senescence by death.

Finally, it is appropriate to return to the question of the meaning of "indirect parental care." Specifically, does this term mean anything other than altruism between individuals of high genetic relatedness? In the rotifer study and in the aphid example cited by Hamilton, altruism and indirect parental care are likely to be identical because both involve clones. "Parental care" as a term may be too restrictive in these situations because of the potential coexistence of many genetically identical individuals having different parents. However, the proposed scheme of life span evolution is expected to pertain to many sexual as well as asexual breeding systems and in these cases parents and their offspring will have much more genetic similarity than parents and individuals randomly drawn from the population. The greater this genetic divergence between parents and average individuals, the more direct should be the parental care. As Alexander (1974) noted, parental care evolves not to increase the fitness of the offspring, but to increase the fitness of the parent. In this context, the term "altruism" seems inappropriate.

Summary

Current theories of aging explain the onset of senescence as being related to the termination of reproduction. At that point direct selection can no longer remove alleles with deleterious effects on fitness. However, these theories do not address the termination of senescence by death, and many organisms have life spans that are substantially longer than their reproductive periods.

It is hypothesized that under certain broad conditions, life span is expected to be twice the length of the generation time. This expectation is generated by proposing beneficial effects from either direct or indirect parental care behavior on the survivorship and fecundity of offspring. These beneficial effects later will be reversed by competitive effects if the parents continue to survive. Senescence provides for a gradual withdrawal from the environment, and a reduction of competition between parents and their offspring. Under this view both senescence and life span are subject to natural selection operating at the level of the individual.

This hypothesis is examined using life table data from 36 treatment groups of a single clone of the rotifer *Brachionus plicatilis*. Treatment groups differed in both chemical composition of culture media and in daily quantity of algal food. Mean survivorship varied from a low of 2.7 days to a high of 11.8 days in these experiments. Over this broad range of environments, life span was found to equal twice the generation time, thus providing support for the proposed hypothesis. It is emphasized, however, that the hypothesis should be considered with continued skepticism until a broader series of tests involving a variety of taxonomic groups has been undertaken.

Acknowledgments. My development of the ideas in this paper was substantially aided by critical discussion with Peter S. Dawson, Paul A. Roberts, members of the Oregon State University Population Biology Discussion Group, and by an anonymous reviewer who made numerous extremely helpful comments. Expert technical help was provided by B. Bayne, S. Dawson, J. Jessup, A. King, L. Lutz, C. Thomas, and A. Weaver. Finally, acknowledgment is due the National Institute of Aging of the U.S. Public Health Services for research funding under grant AG02065.

References

Alexander, R. D.: The evolution of social behavior. Ann. Rev. Syst. Ecol. 5, 325-383 (1974).

Caswell, H.: On the equivalence of maximizing reproductive value and maximizing fitness. Ecology 61, 19-24 (1980).

Charlesworth, B.: Evolution in Age-Structured Populations. Cambridge: Cambridge University Press, 1980.

Cole, L. C.: The population consequences of life history phenomena. Q. Rev. Biol. 29, 103-137 (1954).

Dawson, P. S.: Directional *versus* stabilizing selection for developmental time in natural and laboratory populations of flour beetles. Genetics 80, 773-783 (1975).

De Steven, D.: Clutch size, breeding success, and parental survival in the tree swallows (*Iridoprocne bicolor*). Evolution 34, 278-291 (1980).

Gilbert, J. J.: Mictic female production in the rotifer *Brachionus calyciflorus*. J. Exptl. Zool. 153, 113-124 (1963).

Gilbert, J. J.: Mictic female production in monogonont rotifers. Arch. Hydrobiol. Beih. Ergebn. Limnol. 8, 142-155 (1977).

Greenwood, P. J.: Mating systems, philopatry and dispersal in birds and mammals. Anim. Behav. 28, 1140-1162 (1980).

Hamilton, W. D.: The moulding of senescence by natural selection. J. Theor. Biol. 12, 12-45 (1966).

King, C. E.: Food, age, and the dynamics of a laboratory population of rotifers. Ecology 48, 111-128 (1967).

King, C. E.: Experimental studies of aging in rotifers. Exptl. Gerontol. 4, 63-79 (1969).

King, C. E.: Comparative survivorship and fecundity of mictic and amictic female rotifers. Physiol. Zool. 43, 206-212 (1970).

King, C. E.: Adaptation of rotifers to seasonal variation. Ecology 53, 408-418 (1972).

King, C. E.: Genetics of reproduction, variation, and adaptation in rotifers. Arch. Hydrobiol. Beih. Ergebn. Limnol. 8, 187-201 (1977).

King, C. E.: The genetic structure of zooplankton populations. In: Evolution and Ecology of Zooplankton Communities. Kerfoot, W. C. (ed.). Limnology and Oceanography, Special Symposium 3. Hanover, New Hampshire: University Press of New England, 1980, pp. 315-329.

King, C. E., Miracle, M. R.: A perspective on aging in rotifers. Hydrobiology 73, 13-19 (1980).

King, C. E., Snell, T. W.: Density-dependent sexual reproduction in natural populations of the rotifer *Asplanchna girodi*. Hydrobiology 73, 149-152 (1980).

Lack, D.: The significance of clutch size. Ibis 89, 302-352 (1947).

Lamb, M. J.: Biology of Aging. New York: Halsted Press, Wiley, 1977.

Leslie, P. H.: The intrinsic rate of increase and the overlap of successive generations in a population of guillemots (*Uria aalga* Pont.). J. Anim. Ecol. 35, 291-301 (1966).

Lewontin, R. C.: Selection for colonizing ability. In: The Genetics of Colonizing Species. Baker, H. G., Stebbins, G. L. (eds.). New York: Academic Press, 1965, pp. 77-94.

Lints, F. A.: Genetics and Aging. Interdisciplinary Topics in Gerontology, Vol. 14. Basel: Karger, 1978.

Medawar, P. B.: The Uniqueness of the Individual. London: Methuen, 1957, pp. 44-70.

Rose, M., Charlesworth, B.: A test of evolutionary theories of senescence. Nature (London) 287, 141-142 (1980).

Schaffer, W. M.: Selection for optimal life histories: The effects of age structure. Ecology 55, 291-303 (1974).

Snell, T. W.: Intraspecific competition and population structure in rotifers. Ecology 60, 494-502 (1979).

Snell, T. W., King, C. E.: Life span and fecundity patterns in rotifers: The cost of reproduction. Evolution 31, 882-890 (1977).

Stearns, S. C.: Life history tactics: A review of the ideas. Q. Rev. Biol. 51, 3-47 (1976).

Taylor, H. M., Gourley, R. S., Lawrence, C. E., Kaplan, R. S.: Natural selection of life history attributes: An analytical approach. Theor. Pop. Biol. 5, 104-122 (1974).

Williams, G. C.: Pleiotropy, natural selection, and the evolution of senescence. Evolution 11, 398-411 (1957).

Woolfenden, G. E.: Florida scrubjay helpers at the nest. Auk 92, 1-15 (1975).

Wynne-Edwards, V. C.: Animal Dispersion in Relation to Social Behavior. New York: Hafner, 1962.

Chapter 8

Life History Variation in Dioecious Plant Populations: A Case Study of *Chamaelirium luteum*

THOMAS R. MEAGHER and JANIS J. ANTONOVICS*

Introduction

Perhaps the most commonly acknowledged genetic polymorphism in animal populations is that of sexual dimorphism. Studies on this phenomenon in animal species
have shown that sexual dimorphism influences almost every aspect of their ecology
and evolution (e.g., Bartholemew 1970, Feduccia and Slaughter 1974, Jackson 1970,
Morse 1968). Although perhaps not so conspicuous for plant species, sexual dimorphism can have far-reaching effects on their biology as well. Studies on sexual dimorphism
in plants in the past have been largely limited to floral characteristics (Lloyd and Webb
1977), but over the past few years there has been a growing interest in manifestations
of sexual dimorphism in plant life histories (Grant and Mitton 1979, Hancock and
Bringhurst 1980, Lloyd and Webb 1977, Onyekwelu and Harper 1979, Wallace and
Rundel 1979, Willson 1979). In order to elucidate the role of life histories in the evolution of dioecious plant species, it is necessary to study how sexual dimorphism influences the life histories of males and females and, in turn, how sexual dimorphism in
life history traits, such as mortality and fecundity, influences the relative roles of males
and females in the population as a whole.

The influences of male and female differences on overall life history for plants and
animals share some common principles. Perhaps the most obvious point is that the
female contribution to population dynamics is more conspicuous in the observable
generation of seed or offspring, and so more conducive to measurement. However, by
exclusively emphasizing female contributions, most past studies have only investigated
approximately half (depending on the sex ratio) of the genotypes in the population. It
remains possible that the other half of the genotypes, i.e., the males, may show very
different traits that influence the observed characteristics of the overall populations
(Keyfitz 1977).

Another major drawback of observing only female life history traits is that this
approach is based on the assumption that the dynamics of the female subpopulation is
independent of the male subpopulation, which is clearly not the case for any sexually

*Department of Botany, Duke University, Durham, North Carolina 27706 U.S.A.

reproducing species. For example, female reproductive output is going to depend on the availability of and abundance of the male component of the population. A realistic population projection has to take into account life history traits of both sexes and their interaction. This "problem of the sexes" has been almost the exclusive domain of theoreticians and human demographers (e.g., Das Gupta 1972, 1976, Goodman 1953, 1967, Mitra 1976) and has received little if any attention in the context of natural populations in the zoological and botanical literature.

In most plant species, the impacts of male sexuality and female sexuality on life history are confounded because they take place within the same individuals (cf. Lloyd 1976, 1979b). Dioecious plant species are therefore ideally suited to analysis of the relative impacts of maleness and femaleness on life history because populations of such species consist of separate male and female individuals. Male reproductive contribution can be assayed conveniently by observing whether or not an individual male plant flowers. Thereby it can be determined, for a particular flowering season, which males are most likely to have contributed gametes to the seed crop for that year. If pollen dispersal profiles were known, it would even be possible to assess the relative genetic contribution of particular males on the basis of their proximity to particular females. Even in the absence of such detailed distributional information, the average contribution per flowering male can still readily be assessed. For example, if only a fraction of the males present were in flower in a given season, or if the population sex ratio were shifted toward an excess of females, then the average contribution per male would be increased.

This chapter addresses the contrast between male and female life histories of the dioecious long-lived forest floor herb, *Chamaelirium luteum*. The overall morphology and natural history of *C. luteum* (described in Meagher 1980) make it very amenable to life history studies. Plants occur as basal rosettes with no vegetative spread; therefore individuals readily can be identified. The species has a discrete breeding season, flowering in mid-May; only 10-20% of the individuals in a population flower in a given year. Production of new rosette leaves only occurs once a year in the spring. Moreover, this species shows no sex changes (Meagher 1980), even though the sexes are spatially segregated (Meagher 1980) and there is a strongly male-biased sex ratio (Meagher 1981).

Life history studies on natural populations were conducted in four North Carolina sites designated as Natural Area, Seawell, Silver Hill, and Botanical Garden and described in Meagher (1980). In addition, an analysis was conducted of genetic variation in life history and differences between flowering males and females raised under controlled conditions in the Duke University Phytotron.

Overall, this chapter addresses the following specific questions. What are the genetic differences underlying sexual dimorphism in life history traits in *Chamaelirium luteum*? How are such genetic differences between males and females manifested in the real world under natural conditions? Is there an ecological component to observed sex-specific life history variation in natural populations of *C. luteum*? How are the consequences of the reproductive efforts of male and female individuals reflected in their immediate life history characteristics? What are the effects of observed differences between males and females on long-term population projections? Finally, what are the constraints on life history evolution imposed by male and female sexuality?

Resource Allocation Under Standard Conditions

In order to assess the genetic component of life history variation, a series of plants was raised from seed in the Duke University Phytotron. These plants, representing 30 seeds for each of 30 half-sibships, were taken through a number of flower induction cycles consisting of artificial "summers" and "winters" (see Meagher 1981, for description of the growth conditions) until they flowered. As plants came into flower their sex was recorded and they were either harvested or retained for future analyses.

The sex ratio both within and among half-sibships was found to be 1:1 as of the time when most of the plants had flowered (Table 8-1; see also Meagher 1981), indicating that the sex ratio bias observed among adult plants in natural populations results from secondary ecological and life history effects and not from a bias in the sex ratio among seedlings. However, there were sex-specific patterns observed in the number of induction cycles until plants flowered for the first time, i.e., age at first reproduction (Table 8-2). Males clearly tended to flower earlier in their lives than did females. A similar difference in age at first reproduction has been reported for only one other dioecious plant species (Godley 1976); differences in age at first reproduction of males and females have been noted for a variety of animal species (e.g., Nagel 1979, Ainley and DeMaster 1980). There was also significant heterogeneity (log-likelihood ratio test; Bishop et al. 1975) among half-sibships in age at first reproduction (Natural Area G^2 = 82.3, 45 d.f., p < 0.005; Silver Hill G^2 = 76.3, 45 d.f., p < 0.005; Botanical Garden G^2 = 135.0, 45 d.f., p < 0.005), suggesting that there is genetic variation for age of first reproduction beyond the effect of sex. There have been very few studies that have investigated genetic variation for age at first reproduction, and these few (e.g., Lewontin 1965), in contrast to our findings for *C. luteum*, have indicated very little genetic variation for this trait.

A subsample of 52 male and 52 female plants for 19 half-sibships flowering for the first time following the fourth induction period was harvested and analyzed for dry weight of rosette leaves, rhizome, roots, and reproductive structures. The data were subjected to a crossed partial hierarchical analysis of variance (Brownlee 1960) of sex by half-sibship nested within site of origin. In this analysis, variation among half-sibships within populations and among sexes are both measures of genetic variation, the former a measure of variation from genetic effects other than sex and the latter a measure of genetic variation attributable to sexual dimorphism.

Table 8-1. Pooled half-sibship sex ratios among plants that have flowered in the Duke University Phytotron

Site	Males/females[a]
Natural Area	1.01 (253)
Silver Hill	1.30 (209)
Botanical Garden	1.06 (237)

[a] Sample size in parentheses. None of these sex ratios showed a significant departure from 1:1 when tested using log-likelihood ratio tests (Bishop et al. 1975).

Table 8-2. Average "age" at first reproduction for plants raised under standard conditions.

Site	Males	Females	G^2 (5 d.f.)[a]
Natural Area	3.5	4.1	27.2^b
Silver Hill	4.3	4.6	21.6^b
Botanical Garden	3.0	3.7	25.2^b

[a] Males and females are compared using a log-likelihood ratio (G^2; Bishop et al. 1975).
[b] $p < 0.005$.

Mean values for these characteristics (Table 8-3) showed significant differences between males and females for dry weight of roots and reproductive structures (Table 8-4). Dry weight measurements of roots and reproductive structures also show significant variation among half-sibships, indicating that there exists genetic variation for these characters in addition to the overall sex differences. Rosette leaf number was found to be highly correlated with total dry weight ($r = 0.83$ for males; $r = 0.56$ for females), so that rosette leaf number or size can be used as a convenient assay of total plant size. A significant sex × half-sibship interaction was evident for the total dry weight associated with reproductive structures, indicating that patterns of genetic variation among half-sibships were not the same for the two sexes for this trait. It is interesting to note that among dry weight measurements, the only character that showed this significant interaction was associated with reproduction. The general trend which emerges from these data is that females have a more extensive root structure and a larger inflorescence structure. The most striking difference between males and females is in the quantity of resources that are committed to reproductive structures. The act of producing an inflorescence clearly represents a much greater drain on resources for a female than it does for a male.

One can use the dry weight measurements to assess the relative percentage costs of various plant parts (Figure 8-1). Females devote a significantly greater proportion of their resources to reproductive structures, whereas males have a significantly higher proportion of their resources in the rhizome and rosette leaves (Table 8-4). The percentage of the dry weight allocated to rosette leaves, rhizome, and roots showed significant variability among half-sibships. Hence these patterns of resource allocation are genetically variable for *C. luteum*. Aside from the sex-specific differences there appears to be no genetic variation for allocation to reproduction.

A significant sex × half-sibship interaction was found for percentage dry weight in rhizomes, indicating that patterns of genetic variation for this trait were not the same

Table 8-3. Mean dry weight values for plants harvested in the phytotron experiment[a]

Character	Males	Females
Rosette leaves	2.69	3.03
Rhizome	1.56	1.80
Roots	0.73	1.01
Reproductive structures	0.50	1.62

[a] Results of statistical contrasts on these data are presented in Table 8-4.

Table 8-4. ANOVA test results for mean dry weight and mean percentage allocation to various plant parts in Phytotron experiment

Source	d.f.	F ratios for dry weight[a]				F ratios for % allocation[a]			
		Rosette leaves	Rhizome	Roots	Reproductive structures	Rosette leaves	Rhizome	Roots	Reproductive structures
Sex	1	0.7 n.s.	1.9 n.s.	12.6[b]	123.4[b]	42.5[b]	36.5[b]	0.0 n.s.	154.2[b]
Half-sibships within populations	16	1.8 n.s.	1.5 n.s.	2.3[c]	2.5[d]	2.6[d]	2.1[c]	3.0[b]	1.6 n.s.
Sex × half-sibships within populations	16	1.1 n.s.	0.8 n.s.	1.4 n.s.	1.5[c]	1.4 n.s.	2.0[c]	0.9 n.s.	1.2 n.s.

[a] n.s., not significant.
[b] $p < 0.001$.
[c] $p < 0.05$.
[d] $p < 0.005$.

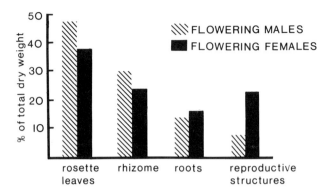

Figure 8-1. Percentage dry weight allocated to various plant parts among flowering males and females raised under standard conditions. Results of statistical contrasts on these data are given in Table 8-4.

for males and females. If the rhizome serves in part as a storage organ for *C. luteum*, which is quite probable, then these differences between the sexes may result from the fact that flowering brings about a much greater resource depletion in females than in males. Under such circumstances, the storage organ would be expected to be subject to differential demands in the two sexes, which in turn might lead to genetic differences.

Resource Allocation and Sexual Dimorphism in Natural Populations

The genetic differences between males and females in their life history characteristics as measured under standard conditions are also strongly evident in natural populations (Table 8-5; see also Meagher 1980, 1981, 1982, Meagher and Antonovics 1982).

Table 8-5. Life history characteristics of male and female plants measured in natural populations[a]

Site	Sex	N	Rosette leaf number	Mortality rates (% per year)	Percentage flowering in a given year
Natural Area	Males	2492	4.1	3.0	34
	Females	1014	4.4[b]	2.6 n.s.[c]	13[b]
Seawell	Males	1025	4.4	1.7	35
	Females	260	4.9[b]	4.0 n.s.[c]	15[b]
Silver Hill	Males	628	4.9	1.3	43
	Females	298	5.3[b]	5.1[d]	14[e]

[a] Values presented are from 1975-1979 data pooled; cumulative numbers of observations (N) for each sex are shown. Statistical significance of male versus female comparisons are indicated by other notes. Rosette leaf numbers were compared using an ANOVA (Sokal and Rohlf 1969); mortality rates and percentage flowering were compared using a log-likelihood ratio (Bishop et al. 1975).
[b] $p < 0.001$.
[c] n.s., not significant.
[d] $p < 0.01$.
[e] $p < 0.05$.

Of the two sexes, females had a significantly higher number of rosette leaves in all three sites, and among plants in flower in a given year, a larger inflorescence as indicated by the number of inflorescence stalk leaves (Meagher and Antonovics 1982). The difference between males and females in annual mortality rate was not statistically significant in the Natural Area but it was so in the other two sites, with females having a higher mortality rate than males. Females therefore tended overall to have a higher mortality rate than males. Finally, females have a lower probability of being in flower in a particular year than do males and they tend to skip a greater number of years after flowering before they flower again (Meagher 1981, Meagher and Antonovics 1982). Consequently, females flower less frequently over a given span of years than do males.

Rosette size (number of rosette leaves) also appears to play an important role in determining the life history characteristics of males, females, and juveniles. Rosette size is highly correlated with inflorescence size for flowering individuals of both sexes and is also a good predictor of the change in size from one year to the next (Meagher and Antonovics 1982). Annual changes in rosette size are sex specific in that females tend to undergo more dramatic fluxes in size than males from year to year (Meagher and Antonovics 1982).

The rosette size of an individual is clearly a good assay of its life history status, as indicated by the correlations found between rosette size and dry weight indicated in the previous section and by the dependence upon size of most life history characters measured in the field. The change in rosette size observed in individuals with different flowering schedules therefore provides a reasonable assay of the drain on a plant's resources brought about by flowering as reflected in the individual's life history. The effects of flowering on rosette size transitions were assessed by considering 3-year size transitions of plants with different flowering schedules. The change in plant size from the year before to the year after flowering was estimated for plants that did not flower in year 1 or year 3 (Table 8-6). Plants did not fully recover their preflowering size by the end of the first year after flowering, and the net loss in size was significantly ($p <$ .05) greater for females than for males at all three sites.

The resource allocation patterns in the phytotron showed that flowering introduced a greater depletion of resources for females than for males. Demographic studies in the field on the impact of flowering on size and subsequent reproduction confirm that the additional drain on female resources has significant impact on the life history of females. In fact the differences between the two sexes have influenced their ecological tolerances such that males and females were differentially distributed into differing

Table 8-6. Percentage change in rosette leaf number from the year before to the year after flowering (year 3 - year 1)[a]

Site	Males		Females	
	n	(year 3 - year 1)	n	(year 3 - year 1)
Natural Area	247	-17	133	-39
Seawell	118	-16	34	-39
Silver Hill	28	+2	12	-34

[a] Statistical comparisons of males and females were all statistically significant ($p < 0.05$), using t-tests (Sokal and Rohlf 1969).

ecological microhabitats within the Natural Area site (Meagher 1980). In the light of life history overall the female life history therefore can be seen as a result of selection favoring female plants that are buffered against this higher cost. In keeping with the predicted outcome of such past selective forces, females delay reproduction presumably until they can obtain a greater rosette size relative to males, providing them with a higher photosynthetic productivity and hence a greater ability to meet and recover from the resource demand of flowering. A tendency toward larger rosette sizes for females among plants in natural populations was also noted. For a given size of inflorescence, larger overall size would also result in a lower proportion of dry weight committed to reproductive structures.

The question remains why females should be constrained to produce a larger inflorescence rather than, for example, producing smaller inflorescences and flowering more often. One possible reason is that seed dispersal may be an important fitness component. Following the flowering season, female inflorescence stalks undergo a secondary elongation to almost double their height by the time seed are shed. There are no specialized mechanisms for seed dispersal (such as fleshy fruits or exploding seed capsules), and this additional elongation seems the only process that enhances seed dispersal in *C. luteum*. If this secondary elongation is a process that has evolved in response to selection for increased seed dispersal distances, then that may explain why females are constrained to produce such large inflorescences.

It has become strongly evident in the preceding sections that males and females are divergent in the manner in which their fitness is expressed through various components (Table 8-7). Males begin flowering younger and have a lower adult mortality rate so that they, on average, have a longer adult life span. Males also have a higher probability of being in flower in a given year. Females, in contrast, by virtue of the fact that they are in a numerical minority and by virtue of the fact that all male gametes passed on to the next generation must go via their seed production, have a much higher number of offspring produced per individual when they are in flower.

Components of Fitness and Population Projection Models

The various components of fitness for males and females can be integrated into a demographic projection model to assess the long-term effects of sexual dimorphism in life history on population dynamics. Such a model is similar in principle to a modification of the Leslie (1945) model whereby the population is subdivided into stage classes rather than age classes (Lefkovitch 1965). The matrix projection model considered here (Figure 8-2) distinguishes among the contributions of the juvenile, male,

Table 8-7. Relative importance of various components of fitness for males and females

Fitness component	Relative importance
Age of first reproduction	Males > females
Adult longevity	Males > females
Probability of flowering	Males > females
Genetic contribution to seedlings when in flower	Females > males

and female components of the population to overall population dynamics and, by extrapolation, to the intrinsic rate of increase of the population, r. A more detailed representation of this model, including size-specific effects, is given in Meagher (1982). For the sake of brevity, the model presented here does not include size-specific effects. There are a few minor discrepancies, therefore, between the presentation here and that in Meagher (1982) in values of r and in proportions of plants in different sex categories resulting from this difference, but the results are generally very similar.

The first row of the matrix in Figure 8-2 represents contributions of the various components of the population at time t to the juvenile component at t + 1 and consists of (1) juvenile survivorship and the portion of the total seedlings attributed to juveniles that flowered for the first time at t, (2) the average contribution to seedlings at t + 1 by males in flower at t multiplied by the probability that any given male would be in flower at time t, and (3) the average contribution to seedlings at t + 1 by females in flower at t multiplied by the probability that any given female would be in flower at t. The male and female fecundities per flowering individual at t are assumed to be represented by half for each sex of the seedlings observed undergoing recruitment at t + 1. The portion of seedlings attributed to juveniles reaching sexual maturity was based on the estimated proportion of plants in flower that were reaching sexual maturity (Meagher 1981, 1982). The second row of the matrix in Figure 8-2 represents contributions of each component at t to the male component at t + 1 and includes (1) the probability that any given juvenile will flower for the first time as a male at t and (2) adult male survivorship. Similarly, the third row consists of (1) the probability that any given plant will flower for the first time as a female at t and (3) female survivorship. The third element of the second row and the second element of the third row represent the probabilities that an individual would undergo a sex change from female to male or male to female, respectively, and they are both equal to zero for this strictly dioecious species.

The differential effect of males and females on long-term projections can perhaps best be assessed by modifying the model to obtain a male-dominance model and a female-dominance model (Keyfitz 1977). This can easily be done for either sex by setting the contribution to seedlings from one sex equal to zero and doubling the contribution to seedlings by the other sex. The effect of this is to subsume the genetic contribution of the second sex into the genetic contribution of the first sex for pur-

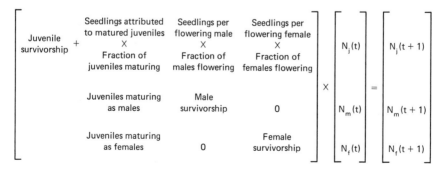

Figure 8-2. General demographic projection matrix for two sexes.

poses of population projection. One-sex projection matrices for *C. luteum* obtained in this manner (Table 8-8) do indicate a difference in the long-term effect of males vs. females. The female-dominance projections generally show a high intrinsic rate of increase than the male-dominance models, although the magnitudes of these differences are not very great.

A problem inherent in the types of projections conducted above is that they fail to take into account the possibility of interaction between males and females in the production of seedlings. Specifically, the relative magnitudes of contribution to seedlings per flowering male (C_m) or per flowering female (C_f) are generally calculated at the particular sex ratios prevailing in the populations at one point in time. If the populations were in a state of disequilibrium with respect to sex ratio at that point or if the sex ratios were undergoing continual perturbation in response to sexual selection, then the sex ratios of the populations will be changing over time as the populations approach equilibrium. Consequently, the true values of C_m and C_f will be different for the equilibrium populations, and the projections based on a C_m and C_f that stay constant over time irrespective of what is happening to the sex ratios will be distorted. In order to overcome this difficulty, a series of interactive models was generated (Table 8-9) in which C_m and C_f were calibrated to the new sex ratio as the sex ratio varied. Because the value of C_f for *C. luteum* appears to remain constant over the range of sex ratios observed in natural populations (Meagher 1980, 1982), C_f is assumed to remain constant, but C_m is free to vary as the sex ratio changes. In these projections, as each new matrix was generated C_m was adjusted to the predicted equilibrium sex ratio. This process was repeated iteratively until the matrices stabilized. Changes in r resulting from calibration of C_m also could influence the relative proportions as well as the overall proportions of juveniles maturing as males or females (Meagher 1982); this effect of proportion of juveniles maturing cannot be accounted for in the present projection because it is dependent on the precise distributions of ages at first reproduction in natural populations for both sexes, which are not yet known. However, estimates of r underwent negligible change in the present simulation, so that the initial estimates of proportions of juveniles maturing as males or females probably represent a reasonable approximation.

Estimates of r obtained using the interactive models differed only slightly from the estimates from male- and female-dominance models. A large departure in value between the estimates of r based on the two one-sex models would be an indication that the populations from which our estimates of C_m and C_f were derived were in a state of disequilibrium with respect to sex ratio. Because the estimates of r obtained by all three methods of projection are of very similar magnitude and indicate a very slow rate of increase in population size over time, the populations of *C. luteum* seem fairly near an equilibrium state with respect to overall life history traits.

Conclusion

There are a number of general properties of forest floor herbs that have become evident in this work on *C. luteum*. Mortality data for all three populations as a whole indicate turnover rates that are relatively low and are hence consistent with observations on other forest floor herbs (Tamm 1956) in contrast to more recent reports of

Table 8-8. Population projections for males and females

Site	Males			Females		
	M	N	r	M	N	r
Natural Area	$\begin{bmatrix} 0.9107 & 0.2859 & 0 \\ 0.0204 & 0.9660 & 0 \\ 0.0190 & 0 & 0.9667 \end{bmatrix}$	$\begin{bmatrix} 57 \\ 22 \\ 21 \end{bmatrix}$	0.019	$\begin{bmatrix} 0.9523 & 0 & 0.4131 \\ 0.0204 & 0.9660 & 0 \\ 0.0190 & 0 & 0.9667 \end{bmatrix}$	$\begin{bmatrix} 67 \\ 17 \\ 16 \end{bmatrix}$	0.047
Seawell	$\begin{bmatrix} 0.9269 & 0.1128 & 0 \\ 0.0247 & 0.9824 & 0 \\ 0.0156 & 0 & 0.9566 \end{bmatrix}$	$\begin{bmatrix} 49 \\ 38 \\ 13 \end{bmatrix}$	0.014	$\begin{bmatrix} 0.9481 & 0 & 0.2750 \\ 0.0247 & 0.9824 & 0 \\ 0.0156 & 0 & 0.9566 \end{bmatrix}$	$\begin{bmatrix} 51 \\ 36 \\ 13 \end{bmatrix}$	0.018
Silver Hill	$\begin{bmatrix} 0.9222 & 0.6651 & 0 \\ 0.0156 & 0.9857 & 0 \\ 0.0192 & 0 & 0.9431 \end{bmatrix}$	$\begin{bmatrix} 73 \\ 15 \\ 12 \end{bmatrix}$	0.059	$\begin{bmatrix} 0.9793 & 0 & 0.5924 \\ 0.0156 & 0.9857 & 0 \\ 0.0192 & 0 & 0.9431 \end{bmatrix}$	$\begin{bmatrix} 75 \\ 14 \\ 11 \end{bmatrix}$	0.067

Table 8-9. Interactive projection matrices and stable population structures

Site	M			N	Interactive model	Males only	Females only
Natural Area	$\begin{bmatrix} 0.9346 & 0.1943 & 0.2066 \\ 0.0204 & 0.9660 & 0 \\ 0.0190 & 0 & 0.9667 \end{bmatrix}$			$\begin{bmatrix} 65 \\ 18 \\ 17 \end{bmatrix}$	0.040	0.019	0.047
Seawell	$\begin{bmatrix} 0.9368 & 0.0469 & 0.1375 \\ 0.0247 & 0.9824 & 0 \\ 0.0156 & 0 & 0.9566 \end{bmatrix}$			$\begin{bmatrix} 48 \\ 39 \\ 13 \end{bmatrix}$	0.013	0.014	0.018
Silver Hill	$\begin{bmatrix} 0.9467 & 0.2206 & 0.2962 \\ 0.0156 & 0.9857 & 0 \\ 0.0192 & 0 & 0.9431 \end{bmatrix}$			$\begin{bmatrix} 71 \\ 17 \\ 12 \end{bmatrix}$	0.050	0.059	0.067

turnover rates for other perennial plant species (see Hickman 1979 for review). However, these more recent studies have been concerned with the dynamics of populations exposed to extreme environmental conditions, such as mine spoils (Antonovics 1972) or ruderal "weedy" situations (Harper 1977, Sarukhan and Harper 1973, Werner 1975, Williams 1970). *Chamaelirium luteum,* however, is an understory herb in mesic mature hardwood forests; and, as such, it is not a transient or colonizing species. Individual plants within populations of *C. luteum* appear to be very long lived and may be as much a lasting part of the community structure of the forest as the trees with which they co-occur.

It is also clear for this dioecious species that differences between males and females in the nature of their contribution to sexual reproduction have far reaching impacts on their life histories (Figure 8-3). The higher "cost" of the female inflorescence stalk results in a much greater resource depletion for females as a consequence of flowering than is evident for males. This higher resource depletion quite probably exposes female plants to a higher risk of mortality as a consequence of flowering, which in turn would result in a higher selection intensity in females for larger plants with more leaf area and hence higher productivity that would help to buffer females against the additional resource depletion that they face. The larger inflorescence in females is believed to have evolved in response to other selection pressures, such as selection favoring increased seed dispersal by females, not measured in the present study. The overall pattern that emerges is that males and females are subject to different internal genetic constraints on resource allocation and that these differences in resource allocation are reflected in a wide spectrum of life history traits.

Even though males and females differed widely in terms of specific life history traits, the long-term projections based on male life history alone (male dominance) and

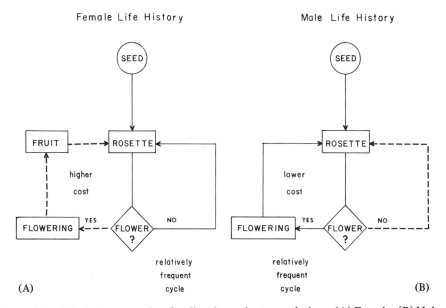

Figure 8-3. Life history overview for dioecious plant populations. (A) Female; (B) Male.

on female life history alone (female dominance) were really quite similar. It seems highly probable that although different components contribute to fitness of males as opposed to females, the net result in terms of overall individual fitness per male and per female are similar. This is not surprising in light of the fact that in any dioecious population the average fitness of the two sexes will be undergoing continual selection toward equality. In studying genetic differences in life histories of contrasting populations or species, these life histories may be the product of different intensities of selection or different constraints within individuals. In the case of dioecy, selection intensities will be similar in males and females because these two sexes make equal contributions to the progeny, and any differences in life history therefore serve to pinpoint the importance of internal constraints on the response to selection for life history traits. The results from the present study show that the constraints can have a large effect on two genotypes within one interbreeding population.

Because of the interdependence of male reproductive output and female reproductive output inherent in sexual reproduction, a population projection should consider the interaction of the two sexes. By studying the life histories of the two sexes separately and developing a projection model that considers relative contributions of both sexes, a more comprehensive understanding of the internal processes and differential nature of the contributions of males and females to overall population dynamics has been gained.

The importance of distinguishing between male and female contributions becomes even more evident in the general case for plants where both male and female functions take place in the same individual but to differing degrees among individuals (e.g., gynodioecy, polygamy; see Lloyd 1980 for discussion of gender states in plants). The relative importance of male and female fitness contributions (within the context of the total population) will determine individual fitness (Lloyd 1976). Furthermore, depending on the magnitude of conflicting constraints of male vs. female reproductive effort, there may be eventual evolution of dioecy (Lloyd 1979a) if, for example the sum of male and female fitness for two individuals of different sex surpasses the sum of male and female fitness of two hermaphroditic individuals. In this latter case, a parent producing dioecious progeny will have greater fitness than one producing hermaphroditic progeny.

The differences in the life history of the two sexes as seen in natural populations was shown in the phytotron studies to have a clear genetic basis. The origins of such a marked genetic dimorphism within a single population are clearly related to the differences in sex function. Even though one would expect strong directional selection for a particular life history that would maximize individual fitness in a population, internal genetic constraints resulting from sexuality of an individual can in fact lead to quite divergent life history variation even within a single population. Dioecy in plants therefore serves as a model system to gain insight into the nature of genetic constraints on life history evolution.

Acknowledgments. The authors would like to thank L. Reinertsen Meagher for her help and support throughout this study, D. G. Lloyd for comments on an earlier draft of this chapter, and members of the Duke Population Genetics group for their comments and suggestions. T. Meagher was supported during this research by a National Institute of Health Graduate Traineeship through NIH grant No. GM02007-08 administered by the University Program in Genetics of Duke University and the latter part of this study by NSF grant No. DEB-7904737. Phytotron studies were supported by NSF grants DEB-7705330 to J. Antonovics for T. Meagher, DEB-7604150 to H. Hellmers for the Duke University Phytotron, and DEB-7904737 to T. Meagher. Travel to field sites was supported by the Graduate School and the Botany Department of Duke University.

References

Ainley, D. G., DeMaster, D. P.: Survival and mortality in a population of Adelie penguins. Ecology 61, 522-530 (1980).

Antonovics, J.: Population dynamics of the grass *Anthoxanthum odoratum* on a zinc mine. J. Ecol. 60, 351-365 (1972).

Bartholemew, G. A.: A model for the evolution of pinniped polygyny. Evolution 24, 546-549 (1970).

Bishop, Y. M. M., Fienberg, S. E., Holland, P. W.: Discrete Multivariate Analysis. Cambridge, Mass.: M.I.T. Press, 1975.

Brownlee, K. A.: Statistical Theory and Methodology in Science and Engineering. New York: John Wiley & Sons, 1960.

Das Gupta, P.: On two-sex models leading to stable populations. Theor. Pop. Biol. 3, 358-375 (1972).

Das Gupta, P.: An interactive nonrandom-mating two-sex model whose intrinsic growth rate lies between one-sex models. Theor. Pop. Biol. 9, 46-57 (1976).

Feduccia, A., Slaughter, B. H.: Sexual dimorphism in skates (Rajidae) and its possible role in differential niche utilization. Evolution 28, 164-168 (1974).

Godley, E. J.: Sex ratio in *Clematis gentianoides* DC. New Zeal. J. Bot. 14, 299-306 (1976).

Goodman, L. A.: Population growth of the sexes. Biometrics 9, 212-225 (1953).

Goodman, L. A.: On the age sex composition of the population that would result from given fertility and mortality conditions. Demography 4, 423-441 (1967).

Grant, M. C., Mitton, J. B.: Elevational gradients in adult sex ratios and sexual differentiation in vegetative growth rates of *Populus tremuloides* Michx. Evolution 33, 914-918 (1979).

Hancock, J. F., Jr., Bringhurst, R. S.: Sexual dimorphism in the strawberry *Fragaria chiloensis*. Evolution 34, 762-768 (1980).

Harper, J. L.: Population Biology of Plants. New York: Academic Press, 1977.

Hickman, J. C.: The basic biology of plant numbers. In: Topics in Plant Population Biology. Solbrig, O. T., Jain, S., Johnson, G. B., Raven, P. H. (eds.). New York: Columbia University Press, 1979, pp. 232-263.

Jackson, J. A.: A quantitative study of foraging ecology of downy woodpeckers. Ecology 51, 318-323 (1970).

Keyfitz, N.: Introduction to the mathematics of population, with revisions. Reading, Mass.: Addison-Wesley, 1977.

Lefkovitch, L. P.: An extension of the use of matrices in population mathematics. Biometrics 21, 1-18 (1965).

Leslie, P. H.: On the use of matrices in certain population mathematics. Biometrika 33, 183-212 (1945).

Lewontin, R.: Selection for colonizing ability. In: The Genetics of Colonizing Species. Baker, H. G., Stebbins, G. L. (eds.). New York: Academic Press, 1965, p. 77-91.

Lloyd, D. G.: The transmission of genes via pollen and ovules in gynodioecious angiosperms. Theor. Pop. Biol. 9, 299-316 (1976).

Lloyd, D. G.: Evolution towards dioecy in heterostylous populations. Plant Syst. Evol. 131, 71-80 (1979 a).

Lloyd, D. G.: Parental strategies of angiosperms. New Zeal. J. Bot. 17, 595-606 (1979 b).

Lloyd, D. G.: Sexual strategies in plants. III. A quantitative method for describing the gender of plants. New Zeal. J. Bot. 18, 103-108 (1980).

Lloyd, D. G., Webb, C. J.: Secondary sex characters in seed plants. Bot. Rev. 43, 177-216 (1977).

Meagher, T. R.: Population biology of *Chamaelirium luteum*, a dioecious lily. I. Spatial distributions of males and females. Evolution 34, 1127-1137 (1980).

Meagher, T. R.: Population biology of *Chamaelirium luteum*, a dioecious lily. II. Mechanisms governing sex ratios. Evolution 35, 557-567 (1981).

Meagher, T. R.: Population biology of *Chamaelirium luteum*, a dioecious member of the lily family. IV. Two-sex population projections and stable population structure. Ecology, in press (1982).

Meagher, T. R., Antonovics, J.: Population biology of *Chamaelirium luteum*, a dioecious member of the lily family. III. Life history studies. Ecology, in press (1982).

Mitra, S.: Effect of adjustment for sex composition in the measurement of fertility on intrinsic rates. Demography 13, 251-257 (1976).

Morse, D. H.: A quantitative study of foraging of male and female spruce-woods warblers. Ecology 49, 779-784 (1968).

Nagel, J. W.: Life history of the ravine salamander (*Plethodon richmondi*) in northeastern Tennessee. Herpetologica 35, 38-43 (1979).

Onyekwelu, S. S., Harper, J. L.: Sex ratio and niche differentiation in spinach (*Spinacia oleracea* L.). Nature (London) 282, 609-611 (1979).

Sarukhan, J., Harper, J. L.: Studies in plant demography: *Ranunculus repens* L., *R. bulbosus* L., and *R. acris* L. I. Population flux and survivorship. J. Ecol. 61, 675-716 (1973).

Sokal, R. R., Rohlf, F. J.: Biometry. San Francisco: W. H. Freeman and Company, 1969.

Tamm, C. O.: Further observations on the survival and flowering of some perennial herbs. Oikos 7, 273-293 (1956).

Wallace, C. S., Rundel, P. W.: Sexual dimorphism and resource allocation in male and female shrubs of *Simmondsia chinensis*. Oecologia 44, 34-39 (1979).

Werner, P. A.: Predictions of fate from rosette size in teasel (*Dipsacus fullonum* L.). Oecologia 20, 197-201 (1975).

Williams, O. B.: Population dynamics of the perennial grasses in Australian semi-arid grassland. J. Ecol. 58, 869-875 (1970).

Willson, M. F.: Sexual selection in plants. Am. Nat. 113, 777-790 (1979).

PART FOUR

Life History Variation Within Populations

Empirical studies of life history traits reveal quasicontinuously variable phenotypes with expression presumably dependent at least in part on polygenic systems. The problem is to relate genotype to phenotype, recognizing that alternative genetic pathways may lead to the same phenotypes and that the same genotypes may be involved in multiple life history phenotypes. Both chapters in this section address this problem by considering samples drawn from single populations.

Doyle and Myers use a set of life history traits in an estuarine amphipod, *Gammarus lawrencianus*, to demonstrate the importance of phenotypic correlations and to illustrate the measurement of selection intensities when such correlations are present. Their study shows that phenotypic correlations modify the intensity of selection acting on life history characters. Hegmann and Dingle use half-sibling analysis to extract genetic correlations from phenotypic correlations among milkweed bug life history traits. Together these two chapters provide substantial empirical evidence of genetic structure relating assemblages of life history components. They reinforce the importance of additive genetic covariance structure for understanding life history evolution, a point also developed from theoretical considerations by Lande's chapter in Part One.

Chapter 9

The Measurement of the Direct and Indirect Intensities of Natural Selection

ROGER W. DOYLE and RANSOM A. MYERS*

Introduction

The intensity of selection on a trait is measured by (or defined by) the effect that the trait has on fitness. Major components of fitness, such as fecundity, will always be selected in a positive direction (Falconer 1960) when the traits are considered in isolation, with other traits held constant. It is obvious, however, that other traits will rarely be constant in populations of real animals. Most traits, especially such composite life history variables as survival and fecundity, are likely to affect fitness in a number of different ways, simply because they do not vary independently in their phenotypic expression. For example, low- and high-fecundity phenotypes may mature at different ages or show differential survival. Correlation of traits at the level of the phenotype is the net result of genetic correlation (pleiotropism, linkage) and environmental variation that influences two or more traits simultaneously (Falconer 1960).

Phenotypic correlations may either reduce or augment the intensity of selection on life history traits. Although life history traits must very often be phenotypically correlated, the effect of this has not received much emphasis in life history theories. Analyses such as those of Caswell (1978) on the relative sensitivity of fitness to changes in development rate and fecundity are strongly influenced by phenotypic correlations, as Caswell himself has pointed out, although a correlation of zero is assumed in his paper.

The purposes of this chapter are (1) to demonstrate empirically that such correlations are important, and (2) to show that the intensity of selection on life history traits can be measured even with the complications of phenotypic correlations included. Selection intensities on several life history traits (fecundity, age at maturation, growth rate, reproductive effort) were measured in an estuarine amphipod, *Gammarus lawrencianus*. We used path analysis to decompose Crow's (1958) index of total selection into the intensity of selection acting directly and indirectly on life history traits. Although the method was used on a species in the laboratory in this case, it can be applied in the field if records are kept on individual organisms.

*Department of Biology, Dalhousie University, Halifax, Nova Scotia B3H 4J1, Canada

Effect of Phenotypic Correlation on Selection Intensity

Natural selection is said to act on a set of traits if variation in the traits causes variation in the survival and reproductive success, or fitness, of individuals in a population. The change in fitness brought about by selection within a generation is (Crow 1958):

$$\Delta W = \frac{V_W}{\overline{W}}$$

and the relative change is:

$$I = \frac{\Delta W}{\overline{W}} = \frac{V_W}{\overline{W}^2} \tag{9-1}$$

where W is the fitness (to be defined below), $\overline{W}$ is the mean fitness before selection, and V_W is the variance in fitness. I is Crow's "index of total selection." This relationship holds true whether the trait is heritable or not.

Following Crow, we subdivide total selection into two components,

$$I = \frac{V_m}{\overline{W}^2} + \frac{1}{P_s} \frac{V_f}{\overline{W}_s^2} = I_m + \frac{I_f}{P_s} \tag{9-2}$$

where the first term on the right is the survivorship component and the second term is the fraction of the total intensity of selection that is attributable to differences in fertility among the survivors. The mean fitness of the population and of the survivors are $\overline{W}$ and $\overline{W}_s$, respectively; V_m and V_f are the variances resulting from differential mortality and fertility, respectively; and P_s is the proportion of the population surviving to reproduction. The mortality component V_m is

$$V_m = (1 - P_s) \overline{W}^2 + P_s (\overline{W}_s - \overline{W})^2$$

Survival and fertility are so-called major components of fitness in the sense that they account for all of the variance in fitness. In this chapter we concentrate on the fertility component of fitness only and make the following notational simplification:

$$\overline{W} \equiv \overline{W}_s$$

and

$$\overline{V} \equiv \overline{V}_f$$

Selection on the fertility component of fitness (henceforth referred to merely as "selection") also can be subdivided into major components: variation in the number of offspring (Darwinian fitness) and variation in the age at which they are produced. This latter component becomes important whenever a population is increasing or decreasing in size, because the relative value of offspring produced t time units in the future differs from the present value of the same number of offspring by the discounting factor exp $(-rt)$ where r is the instantaneous rate of increase of the population. In a growing population, the discount factor of course places a selective premium on early as opposed to late reproduction, other things being equal (Charlesworth 1980).

For a phenotypic trait that is under directional selection, the proportion of the fitness variance attributable to variance in the ith trait can be obtained by simple (not multiple) regression. For example, with a fitness determined by n phenotypic traits,

$$W = f(x_1, x_2, \ldots x_n), \tag{9-3}$$

the component of fertility selection attributable to the ith trait, X_i, is

$$I_i = r^2_{W,i} \frac{V_w}{\overline{W}^2} = a^2_i \frac{V_i}{\overline{W}^2}$$

where $r_{W,i}$ is the simple (not partial) correlation coefficient between W and x_i, a_i is the simple regression coefficient, and V_i is the variance of the ith trait. The increment in fitness attributable to the ith trait is

$$\Delta W_i = a^2_i \frac{V_i}{\overline{W}}$$

The increment in fitness represents a change in the trait x_i of

$$\Delta W_i = a_i \Delta x_i$$

Thus, the selection differential on adults (before reproduction) is

$$\Delta x_i = a_i \frac{V_i}{\overline{W}}$$

This represents the change in the trait x_i in one generation under special circumstances, i.e., if the heritability of the trait is 1.0, if there is no selection on genetically correlated traits, and if dominance and epistasis are unimportant. The above is essentially a restatement of Robertson's (1968) "secondary theorem of natural selection." See Price (1970) for a derivation of a similar expression, Crow and Nagylaki (1976) for the two-locus case, Denniston (1978) for the one-locus case, and Emlen (1980) for applications of Denniston's formalism to the evolution of ecological characters.

The observed numerical value of a_i is strongly influenced by covariance relationships among the variables that are not included in the model shown above. If the variable x_i is negatively correlated with another variable, say x_j, which also affects fitness, the numerical value of a_i is reduced and it may even change sign. The variance in fitness that is such a conspicuous feature of many populations, particularly of marine invertebrates, is not in itself sufficient to demonstrate that these traits are under strong selection. Negative correlations may well reduce the effects of phenotypic variation which, when the traits are considered in isolation, would appear to represent a tremendous loss of fitness in the population. Note that this modification of selection intensity by phenotypic correlations occurs prior to any modification of the response to selection which may be caused by genetic correlation among the traits.

The distinction between direct and indirect selection can be made more formal by distinguishing between the simple and the partial regression coefficients. Constructing a linear fitness function with the first two variables in Eq. (9-3) (now assumed to be phenotypically correlated) and lumping all the other variables together as an error term E (assumed uncorrelated with x_1 or x_2) we have:

$$W = b_{W_1} x_1 + b_{W_2} x_2 + E$$

with

$$r_{12} \neq 0$$

In this constrained fitness function the following relationships hold:

$$\frac{(V_W)_{1 \cdot 2}}{\overline{W}} = \frac{b^2_{W_1} V_1}{\overline{W}} = \Delta W_{1 \cdot 2}$$

where $(V_W)_{1 \cdot 2}$ refers to the direct effect of X_1 on W with X_2 held constant, and b_{W_1} is the partial regression coefficient of W on X_1. Similarly

$$\frac{(V_W)_{2 \cdot 1}}{\overline{W}} = \frac{b^2_{W_2} V_2}{\overline{W}} = \Delta W_{2 \cdot 1}$$

and

$$\frac{(V_W)_{2,1}}{\overline{W}} = \frac{2 b_{W_1} b_{W_2} \, Cov \, (x_1, x_2)}{\overline{W}} = \Delta W_{2 \cdot 1}$$

where $(V_W)_{2 \cdot 1}$ is the variance of fitness attributable to the join effects of x_1 and x_2. The relationship between the simple and partial regressions is straightforward if the fitness function is linear as above:

$$r_{W,1} = \frac{b_{W_1} \sigma_1 + b_{W_2} \sigma_2 r_{12}}{\sigma_W}$$

because

$$a_1 = r_{W,1} \frac{\sigma_W}{\sigma_1} = b_{W_1} + \frac{b_{W_2} \sigma_2 r_{12}}{\sigma_1} \qquad (9\text{-}4)$$

Only if $\text{Cov}(x_1, x_2) = 0$ will the partial be the same as the simple regression coefficient.

The selection differential on x_i is easily expressed in terms of simple and partial regression coefficients:

$$\Delta x_1 = \frac{a_1 V_1}{\overline{W}} = \frac{b_{W_1} V_1}{\overline{W}} + \frac{b_{W_2} \text{Cov}(x_1 x_2)}{\overline{W}}$$

It can be seen that the intensity of selection on x_1 is modified by its covariance with x_2 if the latter is also under selection ($b_{W_2} \neq 0$).

Biology of *Gammarus lawrencianus* in Nature and the Laboratory

Selection for Adaptation to the Laboratory

Gammarus lawrencianus Bousfield is a small euryhaline amphipod that is locally abundant in estuarine habitats from Labrador to Long Island Sound (Steele and Steele 1970). *Gammarus lawrencianus* has been chosen for this study because it grows well in the laboratory and has a relatively short generation time, about 40 days at room temperature. Previous work (Doyle 1978, Doyle and Hunte 1981a) has indicated that it responds rapidly to selection for life history traits and physiological traits, such as resistance to salinity stress.

The population that was used for this experiment was collected from the outlet of Lawrencetown Lake, Nova Scotia (44°38'52" N; 63°21'29" W) in the summer of 1975. This founding population consisted of approximately 100 pregnant females and the same number of mature males. These animals and their descendants were maintained for 5 years in the laboratory under conditions designed to maximize density-independent population growth rate (for procedure see Doyle and Hunte 1981b).

The extent to which life history and demographic traits changed during a 3-year period of adaptation to the laboratory can be seen in Table 9-1. Note that the mean values and the coefficients of variation of the life history traits both changed considerably. We are measuring the intensity of selection on a species previously adapted to a variable (estuarine) environment that has been subject to "natural" selection for survival and growth in a constant laboratory environment.

Table 9-1. Summary of the changes in life history traits after 3 years (approx. 26 generations) under conditions described in the text[a]

Population	45-day survivorship	Age at maturity	All females		Pregnant females		r
			Fecundity at 45 days	RE	Fecundity at 45 days	RE	
Control							
(mean)	0.28	35.5	14.4	1.46	24.4	2.26	0.18
(CV)		17	105	88	50	36	
Laboratory adapted							
(mean)	0.52	28.5	26.6	2.12	31.3	2.37	0.31
(CV)		12	63	52	43	33	
Percentage change in mean	86	−20	85	45	28	4.9	72
Percentage change in CV		−29	−40	−41	−14	−8.3	

[a] CV, coefficient of variation (standard deviation divided by the mean); fecundity, number of eggs in the brood pouch at 45 days; RE (reproductive effort), fecundity per parental wet weight in milligrams; r, intrinsic rate of population increase. Note that here fecundity is measured at 45 days whereas in the rest of this chapter fecundity is measured at maturation.

Definition of Fitness

First Definition:

The fitness of an animal can be defined to be

$$W = \frac{F}{e^{rT}}$$

where F is the number of offspring (counted at the time of first reproduction), T is the age of first reproduction, and r is the instantaneous rate of increase of the population. The value of r is assumed to be constant on the time scale of T. The denominator $\exp(rT)$, which normalizes for the population size at time T and therefore places a selective premium on early reproduction, is what makes this form of selection "r selection" in the literal sense instead of simple fecundity selection.

Second Definition:

The definition of W given above resembles the reproductive value of an individual at age T, but because it is an observed property of each animal the variable l_x is missing. The great majority of animals that survive to breed once also produce subsequent broods and these offspring certainly contribute to the fitness of individual animals, as they do to the rate of population increase. The calculation of r made in this chapter does include the rate of production of postmaturational offspring (see next section),

but broods produced by animals subsequent to their first reproduction (postmaturational reproduction) have not been observable because of the mass rearing techniques employed. The contribution to fitness from postmaturational reproduction can
be estimated, however, by adding an estimated reproductive value term to the directly
observed W defined above. The estimated total fitness is then

$$W^* = W + \sum_{x=T}^{6} e^{-rx} l_x m_x$$

Estimates for the period between broods (9 days), and l_x (age-specific survivorship,
assumed constant at 0.682) were taken from independent experiments. The values of
m_x (age-specific fecundity) were calculated for each female from her weight and
reproductive value at maturation using empirically determined relationships (see next
section). Note that m_x is being used here in a nonstandard manner; it is usually a
demographic or population parameter, whereas here a value is calculated for six
different ages for each surviving female.

The experimental conditions were set up to maximize r by keeping the environmental conditions favorable and the population uncrowded by means of a random
culling procedure (Doyle and Hunte 1981a). The relationship between W and W* for
the population is shown in Figure 9-1. The strong correlation ($r^2 = 0.92$) between the

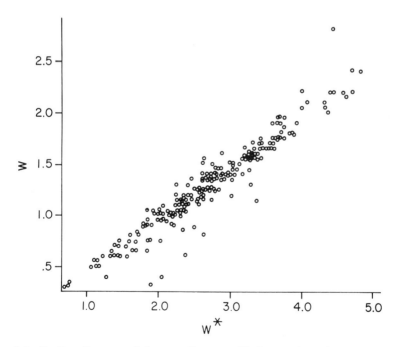

Figure 9-1. Scatter diagram of the two fitnesses, W the number of eggs at maturation
discounted by the instantaneous rate of increase of the population, and W* the expected total discounted egg number over the lifetime of a female, for 213 female
Gammarus lawrencianus ($r^2 = 0.920$).

definitions of fitness is largely a result of the large instantaneous rate of increase. The value for fitness, W*, is the best estimate available for the population. The variance of W and W* will be used to estimate the intensity of selection on several life history traits in the population adapting to a constant laboratory environment. The expected fitness, W*, does not include the effects of all phenotypic correlations in the populations. For example, there is some evidence that growth rate and survivorship are negatively correlated in this species (Doyle and Hunte 1981b), similar to the "Rosa Lee" phenomenon in fish (Ricker 1969). Such correlations may increase or decrease the calculated intensity of selection in the population. The technique of calculating fitness from empirically determined trade-offs is less desirable than following individuals throughout their lives. However, in many situations either such data are impossible to obtain or obtaining them would disturb the organisms to such a degree that they would be useless.

Phenotypic Correlation Among Life History Traits

A cohort of approximately 600 newly hatched juveniles provided most of the data on the correlation among traits. The animals were reared together in a 200-liter tank, and as each female matured (the criterion of maturity was the appearance of fertile eggs in the brood pouch), she was removed from the tank, the eggs were gently removed from the brood pouch and counted, and the animal was weighed. There was essentially 100% survival of females during this procedure. The females were then placed in one of three tanks depending on the ratio of eggs to female wet weight (RE, reproductive effort). Males were added to the tanks so that reproduction could continue. Samples of females were weighed and RE measured every 2 weeks from these tanks to obtain data on postmaturation reproduction. Figure 9-2 shows the growth rate of the groups of females that had high and low reproductive efforts at maturation. The reproductive effort was not significantly different between these two groups 2 weeks after they reached reproductive maturation. (Further experiments are being carried out to see whether this is a repeatable phenomenon with this species.) Data from Figure 9-2 and from the reproductive effort measurements were used to estimate the expected total fitness, W*, for each of the 213 surviving females. The fitness of each female was then calculated from the above equations. The value of r used in the calculation was 0.49/ week, obtained in separate experiments to be reported elsewhere. A cumulative plot of maturation time is shown in Figure 9-3 on a probability scale. The age at maturation is skewed to the right. The distribution of fitness in the population is shown in Figure 9-4.

Estimation of Selection Intensities

To estimate the influence of phenotypic correlations on the selection process it is necessary to separate the effect of each trait on W and W* into two components, the direct effect and the sum of all the indirect effects involving correlation with other traits that also are under selection. A way of making the separation that is both

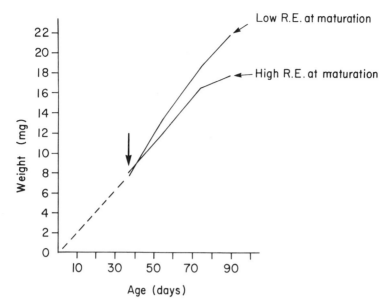

Figure 9-2. Comparisons of the growth of females with low reproductive effort (RE) (number of eggs over weight of female) and high reproductive effort at maturation. The low and the high groups represented approximately 25%, respectively, of the 213 females in the experiment. These and other data were used to estimate W*. The arrow is the average age at maturation.

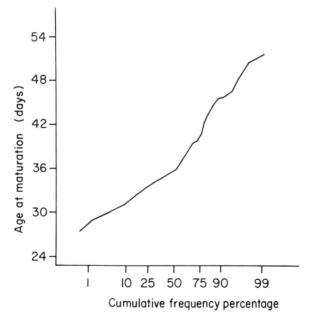

Figure 9-3. The cumulative age at maturation of 213 females plotted on a probability scale. Note that the age of maturation is skewed toward younger individuals.

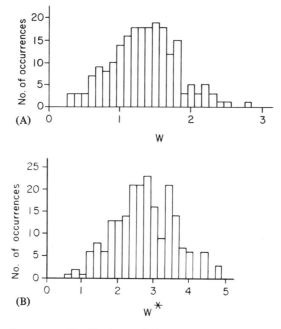

Figure 9-4. The frequency distributions of the two measures of fitness, W (A) and W* (B), in the population.

straightforward and familiar to most population geneticists involves the use of Sewall Wright's path coefficients to describe the relationships among the variables. For an introduction to Wright's papers on the subject and a critique of the method, see Kempthorne (1957).

Path coefficients differ from partial regressions in being "standardized" with all variables expressed in terms of their own population standard deviation. The following simple relationship between path coefficients and partial regression coefficients results:

$$p_{W_1} = b_{W_1} \frac{\sigma_1}{\sigma_W}$$

The determination of fitness variation by variables x_1 and x_2 is

$$p_{W_1}^2 + 2 \, p_{W_1} p_{W_2} r_{12} + p_{W_2}^2 = R^2$$

with

$$R^2 + p_{WE}^2 = 1$$

where R^2, the square of the multiple correlation coefficient, is the proportion of V_W ascribable to all the direct and indirect effects of x_1 and x_2; p_{W_1} is the path coefficient measuring the causal effect of variation in x_1 on V_W and p_{W_2} is defined similarly. The path coefficient p_{WE} represents the effect of all other variables in the system, assumed to be uncorrelated with x_1 and x_2.

By Eq. (9-2) and the above, the proportion of I that is the direct effect of x_1 is:

$$I_1(\text{direct}) = p_{W_1}^2 I/R^2 \tag{9-5a}$$

Similarly, the proportion of I that is directly from x_2 is:

$$I_2(\text{direct}) = p_{W_2}^2 I/R^2 \tag{9-5b}$$

The proportion of I that results from the joint effects of x_1 and x_2 acting through their correlation with each other is:

$$I_{12}(\text{joint}) = 2 p_{W_1} p_{W_2} r_{12} I/R^2 \tag{9-5c}$$

(Note that the fertility component of fitness is represented by I in this chapter.) Expressing selection intensity in terms of its effect of the value of x_i the selection differential on x_i caused by its "direct" and "net" effect on fitness is defined to be:

$$\Delta X_1(\text{direct}) = \frac{p_{W_1} \sigma_W \sigma_1}{\overline{W}} \tag{9-6}$$

and, as in Eq. (9-4):

$$\Delta X_1(\text{net}) = \frac{a_1 V_1}{\overline{W}} \tag{9-7}$$

When one or more intermediate variables intervene between the independent variables and fitness, as, for example, when the effect of RE on W follows a path through F in Figure 9-7b, the overall path coefficient is the product of the intermediate coefficients. The corresponding equation for "net" contribution of x_1 to I is, from Eq. (9-5):

$$I_1(\text{net}) = a_1^2 \frac{V_1}{V_W} \frac{I}{R^2} \tag{9-8}$$

and similarly for I_2.

This interpretation of the meaning of "direct" and "indirect" effect is consistent with Wright's usage but somewhat at variance with that of Kempthorne. In any case we have chosen to use path coefficients instead of ordinary partial regression coefficients because our purpose has been to find out the relative importance of the determining variables and the proportionate modification of direct selection by indirect effects.

The major obstacle to using path coefficients to relate fitness to its component variables is that the technique is based on a linear model and the equations used to define fitness in this chapter are obviously nonlinear. Because W is completely determined by F and T there is no error term and multiple regression of ln W on ln F and T should give a coefficient of determination of 1.00, as indeed it does. When W is regressed on F and T, however, R^2 is 0.91, which indicates that the "error" introduced by the linearity assumption amounts to approximately 9%. A scatter diagram of W on F, T, and growth rate until maturation is shown in Figure 9-5 and one of W* in Figure 9-6.

This nonlinearity does not have the usual meaning of error in a regression model, which requires that errors be normally distributed and uncorrelated with the variables in the system. However, by keeping the analysis linear we can work with W instead of ln W, at the cost of a loss of precision amounting to about 9%.

Observed Selection Intensities

Selection Intensities on Fecundity and Age at Maturation

The relationship between F and T is shown in Figure 9-8. The marked correlation between fecundity and age at maturity has a major effect on the intensity of selection on both traits, as the path analysis presented in Table 9-2 shows. The model on which the calculations giving rise to Table 9-2 is based is shown in Figure 9-7a. The calculations in Table 9-2 show very clearly that the direct influence of both egg number and age at maturity is strongly modified by the effects of phenotypic correlation between these traits. Age at maturity is of course selected in the negative direction in an expanding population, whereas number of offspring is selected in the positive direction. The correlation between these two fitness traits is positive as expressed here, but in "fitness space," where delayed reproduction reduces the fitness of an individual, "development rate" is a more appropriate (but nonlinear) measure of fitness. Development rate is the reciprocal of the age at maturation and the correlation between F and $1/T$ is of course negative. It can be said, therefore, that a negative phenotypic correlation between two fitness traits lessens the intensity of selection on both of them, just as a negative genetic correlation reduces the response to selection of both of them.

The above analysis was repeated for W* (shown in parentheses in Table 9-2). The use of W* instead of W does not affect the above conclusions.

Further Decomposition of the Intensity of Selection

In the previous section the fertility component of fitness was subdivided into only two components, age at maturation and fecundity. However, the nature of the correlation between fecundity and age at maturation can be studied in more detail by incorporating additional phenotypic variables into the causal system, as in Figure 9-7b. In this diagram the three variables that together determine fitness are reproductive effort (RE), growth rate (G), and age at maturity (T). These three phenotypic variables are independent but may be correlated with each other. They influence the fertility compo-

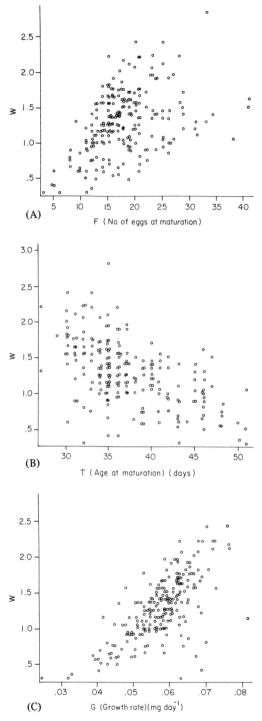

Figure 9-5. Scatter diagram of fitness, W, with (A) F (number of eggs at maturation, $r^2 = 0.320$), (B) T (age at maturation, $r^2 = 0.169$), and (C) G (growth rate to maturation, $r^2 = 0.449$).

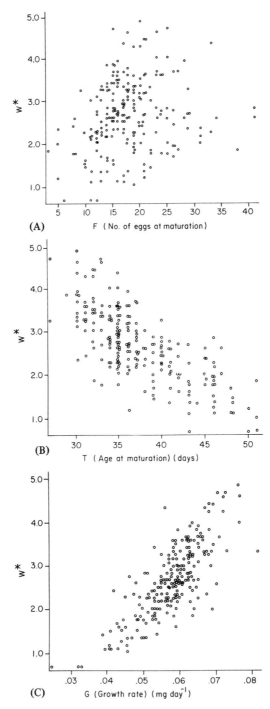

Figure 9-6. Scatter diagram of expected total fitness, W*, with (A) F (number of eggs at maturation, r^2 = 0.040), (B) T (age at maturation, r^2 = 0.536), and (C) G (growth rate to maturation, r^2 = 0.619).

Table 9-2. Intensity of selection on fecundity at maturation (F) and age at maturation (T) in *G. lawrencianus* according to the causal model in Figure 9-7a[a]

Component of fitness	Relationship to fitness	Δx_i	Contribution to I/p_s
F	Direct $(p_{WF})^2$	1.83 (1.32)	0.196* (0.277)
	Net (simple regression)	0.873 (0.390)	0.0448 (0.0241)
T	Direct $(p_{WT})^2$	−0.251 (−0.246)	0.247* (0.639)
	Net (simple regression)	−.147 (−.171)	0.0848 (0.1579)
Joint F,T	$(2p_{WF}p_{WT}r_{FT})$		−0.2047* (−0.3918)
Sum of direct and indirect effects (marked *)			0.2383 (0.2855)
Total	R^2		0.2384 (0.5250)

[a]The selection differential, Δx_i refers to the direct (see Eq. 9-6) and net (Eq. 9-7) components of the selection differential on trait i. The intensity of selection is on the fertility component of fitness only (see Eqs. 9-5 and 9-8). The numbers in parentheses refer to calculations made on W*.

nent of fitness by their effect on one or both of the subcomponents, fecundity and age of maturation. In the diagram it can be seen that size at reproduction, for example, is determined by growth rate and age at reproduction; fecundity is determined by size and reproductive effort, etc. With such a model (supposing that it is a qualitatively correct representation of the real relationships among the variables) it is possible to find out whether independent variables affect fitness mainly through the age at maturation or through fecundity.

The correlation between F and T that was represented by a single double-headed arrow in Figure 9-7a is now resolved into a more complex network of causal paths and correlations among variables not present in the earlier diagram. The path coefficients shown in Figure 9-7b and Table 9-3 were obtained by two-stage least squares (2SLS) (Lesser 1969).

The relative importance of the various routes by which the independent variables affect variation in fitness is shown in Table 9-2. As in the case of the simpler diagram in Figure 9-7a, Wright's line of reasoning was followed in obtaining the partitioning of the relative strength of causal paths. Intensity of selection has been separated into two components, direct (not involving paths going through the other two variables) and net (the resultant of all direct and indirect paths) as explained in the two-variable case. Again, correlations among variables have a profound effect on the intensity of selection, just as in the case of the major components (fecundity, F) previously analyzed.

The most striking observation from Table 3 is the effect of phenotypic correlations on the intensities of selection. The intensity of selection and selection differential were

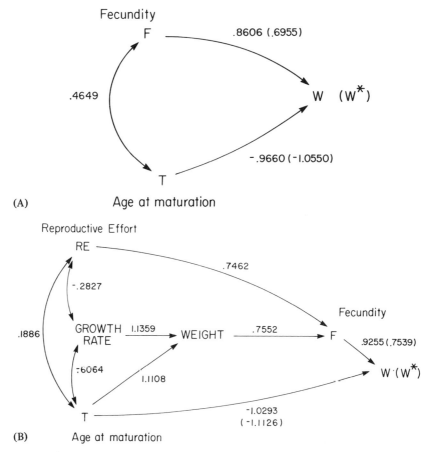

Figure 9-7. Path diagrams of the causal relationships among the life history traits affecting fitness W (path coefficients going to W* are in parentheses). (A) Fitness divided into the two major components, fecundity at maturation and age at maturation (T). This path diagram was used in computing the values in Table 9-2. (B) A more detailed causal model in which the correlation between F and T (represented by a double arrow in Fig. 9-7a) is resolved into a more complex network of paths, including growth rate until maturation, weight at maturation, age at maturation, reproductive effort at maturation, and fecundity at maturation. This causal model was used to obtain Table 9-3.

reduced by phenotypic correlations with reproductive effort and growth rate. Of particular interest is the effect of age at maturation on fitness. There are two direct causal paths (see Fig. 9-7b) linking T with fitness, each having opposite effects on the selection differential.

Again, the use of W* instead of W does not affect the above conclusions.

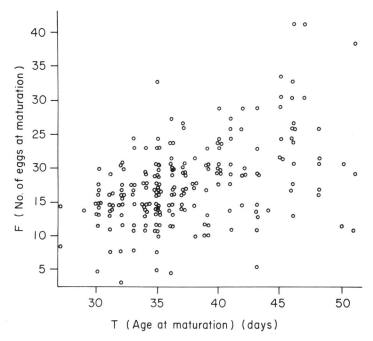

Figure 9-8. Scatter diagram of F (fecundity at maturation) and T (age at maturation) ($r^2 = 0.216$).

Discussion

Alternative Methods of Measuring the Intensity of Selection

The intensity of selection on life history traits must be measured if life history theories are to be tested. The decomposition of Crow's index of total selection provides a method for making such measurements. Wade (1979) and Wade and Arnold (1980) have applied Crow's index to describing the intensity of sexual selection. Their procedure resembles Crow's in being a decomposition of variance components. An alternative approach to ours would be to use concrete path coefficients instead of the standardized ones that we used (Li 1975, Tukey 1954). If the intensity of selection on life history traits were compared in two different environments, the appropriate method would be concrete path coefficients (Li 1975). However, standardized path coefficients determine the degree of impact one variable has on another and they essentially tell the degree of determination of the variance of the dependent variable by another. For decomposing the intensity of selection in one population, therefore, standardized coefficients are preferable, but if the selection intensities on life history traits are compared between populations, the concrete path coefficients are preferable.

Haldane (1954) introduced a measurement of the intensity of selection for a phenotype in terms of the difference in survival rates of the optimum vs. the average phenotype (see also O'Donald 1968, Van Valen 1965). This method is limited to situations

Table 9-3. Intensity of selection on reproductive effort (RE, fecundity over body weight at maturation), growth rate until maturation (G), and age at maturity (T) in *G. lawrencianus* according to the model in Figure 9-7b[a]

Phenotypic variable	Relationship to fitness	Δx_i	Contribution to I/p_s
RE	Direct $(p_{WF}p_{F\ RE})^2$	0.138	0.126*
	Net (simple regression)	0.0908	0.0545
G	Direct $(p_{WF}p_{F\ Wt}p_{Wt\ G})^2$	0.00215	0.1691*
	Net (simple regression)	0.00179	0.117
T	Direct $(p_{W\ T})^2$	−0.289	0.280
	Direct $(p_{WF}p_{FW}p_{Wt\ T})^2$	0.2014	0.159
	Resultant of direct effects	−0.0876	0.0171*
	Net (simple regression)	−0.147	0.0848
Joint RE, growth			0.0827*
Joint RE, T			−0.0175*
Joint T, growth			0.0650*
Sum of direct and indirect components (marked *)			0.278
Total R^2			0.238

[a] Δx_i is the selection differential on trait i (see Eqs. 9-6 and 9-7). I/p_s is the selection intensity on the fertility component of fitness (see Eqs. 9-5 and 9-8).

in which the optimum phenotype can be determined and therefore is more appropriate for the study of stabilizing selection rather than the directional selection considered here. Lande (1979 and Chapter 2, this volume) has recently introduced a new approach to multivariate evolution which takes into account the net selection intensity problem that is the motivation for this chapter.

Testing Life History Theories

Phenotypic correlations may significantly alter the selection intensities on life history traits. For example, the net selection on a trait may be zero even though the direct selection is positive, or the direction of selection may be altered. The method described in this chapter measures the selection intensities on a population at the time of observation but does not describe the selection pressures responsible for the evolution of a trait as Lande (1979) has attempted to do. Direct selection pressures on a population may change over time and even if they do not the phenotypic correlations may change in such a manner that the net selection pressures are changed. Evidence of this can be seen in the 26-generation selection experiment referred to earlier (see Table 9-1 and Doyle and Hunte 1981b). There were large changes in the coefficients of variation in the life history traits that could well mean a change in the phenotypic correlations between these traits.

Few life history studies have attempted to determine the intensities of selection on life history traits. The problems of identification and estimation common to studies of life histories are well known to other disciplines, particularly econometrics (Goldberger 1964). The type of data needed for estimations of the selection intensities are often collected during ecological investigations where records are kept on individuals. As shown in this study, fitness variation sometimes can be estimated from incomplete data if enough is known about the empirical trade-offs between different life history traits. The necessary minimum data are the estimation of the variability in a population during some period of the life history. In this case the period was the age at maturation of a large cohort.

The method presented here is appropriate if the fitness function is approximately linear, i.e., for populations subject to directional selection and not stabilizing selection. In the neighborhood of a local maximum of a fitness function fitness cannot be approximated by a linear function. In such situations, any methods relying on relationships that are approximately linear, such as path analysis, will be useless. Perhaps the most suitable method of studying the evolution of life histories is to study a population undergoing adaptation, either in the laboratory, as was done here, or in "natural experiments" (as when a new species is introduced to an area) or by field manipulations, for example, transplant studies (Law 1979).

Summary

The intensities of selection on life history traits (fecundity, age of maturation, growth rate, reproductive effort) were measured in a population of the amphipod *Gammarus lawrencianus*. The method used to measure the intensity of selection was a decomposition of Crow's index of total selection into direct (not involving other traits), and indirect (resulting from selection on phenotypically correlated traits) causal paths connecting the traits with fitness. Phenotypic correlations were demonstrated to be of great importance in modifying the intensity of selection on life history traits.

Acknowledgments. This work was supported by an operating grant from the Natural Sciences and Engineering Research Council of Canada. We thank Russell Lande for a useful comment on the effect of phenotypic correlation on his selection gradient operator.

References

Caswell, H.: A general formulation for the sensitivity of population growth rates to change in life history parameters. Theor. Pop. Biol. 14, 215-230 (1978).

Charlesworth, B.: Evolution in age-structured populations. Cambridge: Cambridge University Press, 1980.

Crow, J. F.: Some possibilities for measuring selection intensities in man. Hum. Biol. 30, 1-13 (1958).

Crow, J. F., Nagylaki, T.: The rate of change of a character correlated with fitness. Am. Nat. 110, 207-213 (1976).

Denniston, C.: An incorrect definition of fitness revisited. Ann. Hum. Genet. London 42, 77-85 (1978).

Doyle, R. W.: Ecological, physiological and genetic analysis of acute osmotic stress. In: Marine Organisms: Genetics, Ecology and Evolution. Battaglia, B., Beardmore, J. A. (eds.). New York: Plenum Press, 1978, pp. 275-287.

Doyle, R. W., Hunte, W.: Genetic changes in the components "fitness" and yield of a crustacean population in a controlled environment. J. Exptl. Mar. Biol. Ecol. 52, 147-156 (1981 a).

Doyle, R. W., Hunte, W.: Demography of an estuarine amphipod (*Gammarus lawrencianus*) experimentally selected for high "r": A model of the genetic effects of environmental change. Can. J. Fish. Aquat. Sci. 38, 1120-1127 (1981 b).

Emlen, J. M.: A phenotypic model for the evolution for ecological characters. Theor. Pop. Biol. 17, 190-200 (1980).

Falconer, D. S.: Introduction to Quantitative Genetics. New York: Ronald Press, 1960.

Goldberger, A. S.: Econometric Theory. New York: Wiley, 1964.

Haldane, J. B. S.: The measurement of natural selection. Caryologia 6 (Suppl. 1), 480 (1954).

Kempthorne, O.: An Introduction to Genetic Statistics. New York: Wiley, 1957.

Lande, R.: Quantitative genetic analysis of multivariate evolution, applied to brain: body size allometry. Evolution 33, 402-416 (1979).

Law, R.: The costs of reproduction in annual meadow grass. Am. Nat. 113, 3-16 (1979).

Lesser, C. E. V.: Econometric Techniques and Problems. London: Griffin, 1969.

Li, C. C.: Path Analysis: A Primer. Pacific Grove, Calif.: Boxwood Press, 1975.

O'Donald, P.: Measuring the intensity of natural selection. Nature (London) 220, 197-198 (1968).

Price, G. R.: Selection and covariance. Nature (London) 227, 520-521 (1970).

Ricker, W. E.: Effects of size-selective mortality and sampling bias on estimates of growth, mortality, production, and yield. J. Fish. Res. Bd. Can. 26: 479-541 (1969).

Robertson, A.: The spectrum of genetic variation. In: Population Biology and Evolution. Lewontin, R. C. (ed.). Syracuse: Syracuse University Press, 1968.

Steele, D. H., Steele, V. J.: The biology of *Gammarus* (Crustacea; Amphipoda) in the northwestern Atlantic. IV. *Gammarus lawrencianus* Bousfield. Can. J. Zool. 48, 1261-1267 (1970).

Tukey, J. W.: Causation, regression and path analysis. In: Statistics and Mathematics in Biology. Kempthorne, O., Bancroft, T. A., Gowen, J. W., Lush, J. L. (eds.). Ames, Iowa: Iowa State University Press, 1954, pp. 35-66.

Van Valen, L.: Selection in natural populations. III. Evolution 19, 514-558 (1965).

Wade, M. J.: Sexual selection and variance in reproductive success. Am. Nat. 114, 742-746 (1979).

Wade, M. J., Arnold, S. J.: The intensity of sexual selection in relation to male sexual behaviour, female choice, and sperm precedence. Anim. Behav. 28, 446-461 (1980).

Chapter 10

Phenotypic and Genetic Covariance Structure in Milkweed Bug Life History Traits

Joseph P. Hegmann and Hugh Dingle*

Introduction

Variation in life history characteristics is critical for any species because it influences the timing and extent of population growth. When the variability is caused by gene differences among individuals, changes in population growth differ among genotypes. Gene frequencies will change at loci with effects on life history traits, and selection and microevolution of life history traits should occur. However, if many life history characteristics vary among individuals, there may be pairwise covariance caused, in part, by genetic linkage disequilibrium or by pleiotropy. Change caused by selection acting on additive genetic variance for any single character will be accompanied by change in other genetically correlated characters. Darwin (1859; see Darwin 1958) commented on this situation in reference to domestication and artificial selection, which he said "will almost certainly modify unintentionally other parts of the structure, owing to the mysterious laws of correlation" (p. 35). The importance of genetic correlations in the context of life history variables and natural selection is the primary motivation for the work reported here.

The nature of gene effects on life history variation has received a great deal of theoretical (cf. Hairston et al. 1970, Lande, Chapter 2, this volume) and empirical attention (cf. Istock 1978, McLaren 1976). Additive genetic variance for major components of fitness (Robertson 1955) is expected to be low (cf. Fisher 1958, Roberts 1967) because directional selection tends to eliminate this source of variance. Further, additive genetic covariance among major components of fitness should be negative (cf. Dickerson 1955, Falconer 1960, Robertson 1955) as long as the population is encountering environmental pressures similar to those that have acted on its ancestors. This expectation follows from the fact that selection will act, by definition, on genetic variation for major fitness components, and the response to selection will be governed by the additive genetic variance-covariance relationships among the characters (see Lande 1979, p. 406). For example, response to selection favoring increase in expression

*Program in Evolutionary Ecology and Behavior, Department of Zoology, University of Iowa, Iowa City, Iowa 52242 U.S.A.

of one life history trait may be countered by response to selection tending to increase another with which it is negatively genetically correlated. Response of either character to selection in this case, and in any case where the direction of selection on two characters opposes the sign of their genetic correlation, results in "genetic slippage" (Dickerson 1955). When the directions of selection and the sign of the genetic correlation among traits match, however, the direct response to selection may be augmented by indirect response acting through genetic covariation among traits (cf. Falconer 1960). Therefore, equilibrium populations should show negative genetic correlations among fitness components, whereas populations moving to a new adaptive zone (see Lande 1980, p. 470) temporarily may show unexpectedly rapid responses to selection, probably because of shifts in the directions or intensities of selection pressures acting on components of fitness.

Analysis of the genetic basis of variation in life history traits is especially important because of the central importance of the traits to natural populations. Genes affecting life history traits should play a dynamic role in adaptation and evolution. The traits can be expected, theoretically, to be only slightly heritable (if at all) and genetic correlations among them should be maintained (if at all) by the selective trade-offs described by Dickerson (1955). Empirical evaluation of additive genetic variance and covariance for these traits requires a powerful design and large sample sizes because additive genetic variance is likely to be low. The evaluation should be undertaken in animals where environmental influences on life history traits are known so that major adaptive challenges can be avoided when natural populations are reared in the laboratory. Finally, the organism chosen for evaluation should be widely distributed geographically because the effects of adaptation on genetic structure of life history traits is a premier problem in population biology and is best approached using the comparative method (cf. Bell 1980).

Materials and Methods

For these reasons, we undertook a large breeding experiment designed to examine the additive genetic variance-covariance structure influencing a set of life history variables in the milkweed bug, *Oncopeltus fasciatus*. Dingle (1968, Dingle et al. 1980a,b) had previously developed efficient culturing procedures for the bugs and had established a large body of information regarding its behavioral ecology and life history characteristics (including their dependence on photoperiod and temperature). In addition, we had some past experience with genetic analyses in the milkweed bug *Lygaeus kalmii* (Caldwell and Hegmann 1969) and had jointly examined the genetic variance for photoperiodically induced diapause in *Oncopeltus fasciatus* (Dingle et al. 1977). The animal is extensively geographically differentiated with regard to both body size and diapause characteristics (see Dingle et al. 1980a, b and Chapter 12, this volume), and it displays considerable variation among individuals within regions. Therefore the potential for analysis of a single population was good, and the potential to extend the findings in a comparative study was great.

We chose a set of life history characteristics (listed in Table 10-2) that could be measured on every individual female in the study and that related directly to reproduction in the bug. Body size was indexed by both wing length and body length (measured with an ocular micrometer). Age at first reproduction was determined by placing 4-day-old females with males and recording the age at which their first clutch of eggs was deposited. Once mating had begun among individuals, we determined the number of eggs present in each of the first three clutches deposited and the hatching success of the eggs in each clutch. We expected these latter six variables to be positively phenotypically correlated. We had no information on which to base expectations regarding their genetic associations. The final two characters measured were the interval between successive clutches (interclutch interval), and the time required after hatching for individual development to adult eclosion (developmental time). All animals were maintained in LD 16:8 photoperiod and at 23°C with food and water constantly available. They were reared as clutches (full siblings) in plastic containers, the size of which was varied to maintain approximate equal density at each instar for all clutches.

In order to gain estimates of additive genetic variance for each of these characters and of the additive genetic covariances among them which were as free as possible of dominance and epistatic components of variance and covariance, we employed half-sibling comparisons (see Falconer 1960). Specifically, a number of males (sires) were each mated to a number of females (dams) so that both full-sibling and half-sibling families were produced. All measurements discussed in the previous paragraph were determined for a sample of female offspring from each family. Males unrelated to the females being measured were used for breedings.

Because of the limits of culture space and the time requirements for data collection, the experiment was run in three blocks. Eleven males and 37 females one generation removed from bugs collected in the field in Iowa during the summer of 1975 eclosed on about December 1 and gave rise to 401 female offspring which comprised the first block. The second block was similarly generated from 10 males and 37 females from a 1976 summer collection, who gave rise to 446 females. The last set was 314 females produced by 10 males and 30 females, one generation removed from a 1977 summer field collection. The details of the family structure for the experiment are presented in Table 10-1 which indicates the number of female offspring from each pair (for example, dam A under sire 1 in block 1 produces 10 female offspring sampled, dam B produced 14, etc.). The blocked design of our sampling originally promised some information on stability of structure over time, but subsequent analysis clearly indicated that too few families were available in any single block to evaluate the genetic structure of the population. All analyses discussed here are based on the entire data set (11 life history characters from each of 1161 female offspring from 31 paternal half-sibling and 104 full-sibling families).

We used generalized methods for analysis of variance (Finn 1974) to estimate variance components among sires, among dams within sires, and among progeny within dams. We extended the analysis for estimation of covariance components associated with each level of the hierarchical design, following Dickerson (1959). In the half-sibling design employed, the component of variance among sires estimates one-quarter of the additive genetic variance influencing each trait. Likewise, the among-sires component of covariance estimates one-quarter of the additive genetic covariance among pairs of traits. Heritabilites of the life history characters were estimated from the intra-class correlations among half-siblings (see Falconer 1960, p. 175). Genetic correlations were solved directly as the ratio of the additive genetic covariance between traits (from their among-sires covariance component) to the product of the additive genetic

Table 10-1. Numbers of female offspring sampled from families of 31 sires and 104 dams measured in three blocks

	Block 1, N = 401 Sires										
	1	2	3	4	5	6	7	8	9	10	11
	A-10	A-11	A-12	A-12	A-26	A-11	A-11	A-14	A-10	A-4	A-8
Dams	B-14	B-12	B-9	B-13	B-11	B-13	B-13	B-10	B-12	B-7	B-12
	C-12		C-10	C-13	C-4	C-11	C-13	C-8		C-12	C-5
			D-12		D-10		D-12	D-12		D-13	
					E-11						

	Block 2, N = 446 Sires									
	1	2	3	4	5	6	7	8	9	10
	A-11	A-13	A-13	A-14	A-13	A-14	A-14	A-14	A-9	A-14
	B-13	B-12	B-3	B-6	B-12	B-13	B-13	B-13	B-12	B-10
Dams	C-13	C-11		C-14	C-13	C-12	C-5	C-14	C-12	C-13
	D-11	D-14			D-13	D-12	D-14	D-13	D-12	D-14

	Block 3, N = 314 Sires									
	1	2	3	4	5	6	7	8	9	10
	A-12	A-12	A-4	A-10	A-12	A-12	A-11	A-13	A-14	A-9
	B-12			B-14	B-12	B-9	B-5	B-11	B-6	
Dams	C-10			C-9	C-14	C-11	C-12	C-13	C-11	
	D-13			D-9		D-10	D-10	D-8	D-6	

standard deviations for each (from their among-sire variance components). Standard errors of estimates were approximated following Osborne and Patterson (1952) and Falconer (1960).

Results

The means for the life history characters are presented in Table 10-2 to provide an overall orientation to the nature of the variables observed in this experiment. These means can be viewed as a mean vector of quantitative life history characters for the population sampled [see Lande, Chapter 2, this volume, Equation (2-1) and ensuing discussion]. Ten of the 11 characters showed significant differences among half-sibling family groups; the single exception was interclutch interval. Heritability estimates for the traits (also Table 10-2) indicate their potential, as single variables, for response to selection. The effects selection for any of them would have on the multivariate vector of means (Table 10-2) depend on the extent to which they are correlated and degree to which their correlations depend on additive genetic covariance among them.

The direction and magnitude of phenotypic correlations among life history characters is illustrated in Figure 10-1, using a three-dimensional projection in which a pair

Table 10-2. Overall means and heritabilities (± standard errors) for life history parameters of *Oncopeltus fasciatus*

	Mean	Heritability
Wing length (mm)	13.0 ± 0.18	0.55 ± 0.22
Body length (mm)	13.7 ± 0.22	0.20 ± 0.14
Age at 1st reproduction (α) (days)	27.5 ± 0.41	0.25 ± 0.12
First clutch size (eggs)	35.7 ± 0.47	0.25 ± 0.10
Second clutch size	30.4 ± 0.43	0.10 ± 0.07
Third clutch size	27.2 ± 0.39	0.08 ± 0.06
Percentage hatch first clutch	0.77 ± 0.009	0.25 ± 0.12
Percentage hatch second clutch	0.80 ± 0.009	0.23 ± 0.11
Percentage hatch third clutch	0.81 ± 0.008	0.24 ± 0.12
Interclutch interval (days)	2.4 ± 0.06	0.04 ± 0.04
Developmental time (days)	42.9 ± 0.10	0.89 ± 0.32

of x and y axis values specify a pair of characters and the z axis indicates the value of their correlations. Genetic correlations among traits are displayed in the same form in Figure 10-2.

The two indices of body size used in the study were positively correlated both phenotypically (Fig. 10-1 and Table 10-3, upper triangle) and genetically (Fig. 10-2 and Table 10-3, lower triangle). The heritability estimates for these variables suggest that the population sampled would respond to directional selection based on extreme body size (especially wing length) and the high positive genetic correlation among the

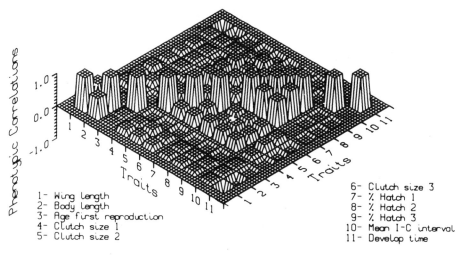

1- Wing Length
2- Body Length
3- Age first reproduction
4- Clutch size 1
5- Clutch size 2
6- Clutch size 3
7- % Hatch 1
8- % Hatch 2
9- % Hatch 3
10- Mean I-C interval
11- Develop time

Figure 10-1. The phenotypic correlational structure influencing 11 life history characteristics of *Oncopeltus fasciatus* descended from an Iowa sample. The viewpoint for plotting is from above and perpendicular to the leading diagonal, which is a row of 11 ones, indicating the correlation of each trait with itself. The figure is plotted with symmetry and the z-axis values, phenotypic correlations, are those in the upper triangle of Table 10-3.

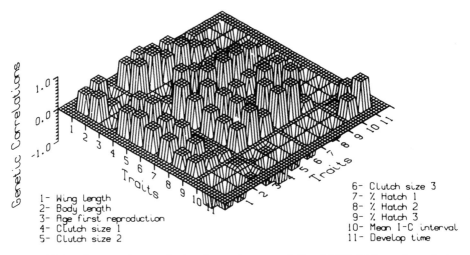

Figure 10-2. The genetic correlational structure associating 11 life history characteristics of *Oncopeltus fasciatus* descended from an Iowa sample. The viewpoint is from above and perpendicular to the leading diagonal, which contains heritability estimates (see Table 10-2) for the traits. Plotted with symmetry and using z-axis values of the genetic correlations detailed in the lower triangle of Table 10-3.

two estimators of size clearly predict that both would respond to selection based on either one. The size of *Oncopeltus* is not significantly associated with age at first reproduction for the bugs, although there is large and significant phenotypic and genetic correlation of the size measures with clutch size and percentage hatch in each of the first three clutches. Size is negatively associated with developmental time, and that negative association has a strong genetic component (genetic correlations of −0.55 and −0.56 between developmental time and the two measures of size). The strong negative relationship between size and developmental time in *Oncopeltus* may indicate that a size threshold controls adult emergence (cf. Blakley and Goodner 1978) and that the threshold varies among families in direct relationship to developmental rate.

Age at first reproduction is influenced by sufficient heritable variation to allow selection response by the Iowa population sampled, and it is notable for its lack of genetic correlation with any of the other life history variables. *Oncopeltus* is known to extend age at first reproduction and enter migratory flight on sufficient environmental cues (Dingle 1978). We have also demonstrated that laboratory populations of *Oncopeltus* maintained in photoperiodic conditions that extend age at first reproduction very quickly show a decrease in age at first reproduction when held in mass culture (Dingle et al. 1977). The relative independence of genetic variance influencing age at first reproduction is consistant with the hypothesis offered in the previous study; namely, that photoperiodically induced diapause serves to "provide optimal association of diapause, migration, and photoperiod in the face of varying association of photoperiod with temperature and food" (Dingle et al. 1977 p. 1055). Strong genetic correlations between age at first reproduction and the other life history characters studied here would tend to "tie up" the genetic variance serving to provide that association.

Table 10-3. Phenotypic (upper triangle) and genetic (lower triangle) correlations among life history characteristics of the milkweed bug, *Oncopeltus fasciatus*

	Wing length	Body length	α	Clutch size			Percentage hatch			X̄ ICI	Development time
				I	II	III	I	II	III		
WL		0.59	0.02	0.15	0.16	0.14	0.09	0.10	0.08	0.05	−0.39
BL	0.67 ± 0.21		0.03	0.10	0.15	0.09	0.05	0.06	0.04	0.05	−0.22
α	−0.29 ± 0.28	−0.41 ± 0.34		−0.09	−0.09	−0.12	−0.08	−0.08	−0.09	0.04	−0.09
I	0.73 ± 0.13	0.91 ± 0.06	−0.22 ± 0.29		0.27	0.29	0.37	0.32	0.30	0.09	−0.10
II	0.80 ± 0.13	1.20 ±[a]	−0.27 ± 0.37	1.09 ±[a]		0.29	0.20	0.26	0.23	0.20	−0.12
III	0.66 ± 0.24	0.69 ± 0.29	−0.98 ± 0.02	0.79 ± 0.16	1.04 ±[a]		0.20	0.21	0.24	0.15	−0.12
I	0.60 ± 0.21	0.77 ± 0.17	0.07 ± 0.34	0.71 ± 0.16	1.06 ±[a]	0.45 ± 0.38		0.80	0.73	−0.01	−0.06
II	0.37 ± 0.27	0.50 ± 0.31	0.11 ± 0.33	0.62 ± 0.19	1.01 ±[a]	0.53 ± 0.33	1.02 ±[a]		0.82	−0.06	−0.07
III	0.38 ± 0.26	0.27 ± 0.37	0.18 ± 0.32	0.63 ± 0.18	0.95 ± 0.04	0.56 ± 0.31	1.01 ±[a]	1.01 ±[a]		−0.07	−0.06
ICI	−0.74 ± 0.21	−0.57 ± 0.41	0.07 ± 0.49	−0.21 ± 0.43	−0.08 ± 0.59	−0.82 ± 0.22	−0.79 ± 0.19	−0.38 ± 0.43	−0.54 ± 0.35		−0.01
DT	−0.55 ± 0.19	−0.56 ± 0.24	−0.13 ± 0.28	−0.57 ± 0.18	−0.76 ± 0.15	−0.58 ± 0.11	−0.52 ± 0.22	−0.33 ± 0.26	−0.44 ± 0.23	0.51 ± 0.32	

[a]Standard errors undefined.

Measures of clutch size and percentage hatch are positively associated phenotypically and genetically. In fact, the extremely high genetic correlations among consecutive clutch sizes suggest that virtually all of the genetic variance influencing them has common affect on all clutches. The same interpretation applies to the three measures of hatchability. In addition, there is strong genetic correlation among measures of clutch size and hatchability, so that selection acting (through the heritabilities shown in Table 10-2) to change any of this set of measures would result in a correlated response for each of the others. This complex of clutch size and hatchability is, of course, positively genetically correlated to body size but negatively genetically correlated to developmental time.

Conclusions

Thus, a complex picture emerges when information from the genetic correlation matrix (lower triangle of Table 10-3) and the vector of heritabilities (Table 10-2) are combined. The population sampled maintains sufficient gene-based variability to show rapid response to directional selection for either body size or developmental time, but progress in response for either must involve a change in the other. Age at first reproduction appears genetically variable, and the character does not seem to be genetically correlated in any systematic way with the other measures of life history dynamics used in this study. Genetic variance for clutch size and percentage hatch of clutches seems sufficient to allow response to selection, and all measures would respond together if selection were to act on any one of the set. However, developmental time would be affected too, in the opposite direction to the correlated effect predicted for body size (because of the negative genetic correlations across the bottom row of Table 10-3).

This study demonstrates that genetic variance for characters closely related to fitness can be detected in empirical studies with sample sizes which are (barely) reasonable. It opens the door for future comparative studies of the genetic variance-covariance relationships among life history characters that are required to assess the generality and stability of the associations observed.

References

Bell, G.: The costs of reproduction and their consequences. Am. Nat. 116, 45-76 (1980).

Blakley, N., Goodner, S. R.: Size-dependent timing of metamorphosis in milkweed bugs (*Oncopeltus*) and its life history implications. Biol. Bull. 155, 499-510 (1978).

Caldwell, R. L., Hegmann, J. P.: Heritability of flight duration in the milkweed bug *Lygaeus kalmii*. Nature (London) 223, 91-92 (1969).

Darwin, C. The Origin of Species by Natural Selection or the Preservation of Favoured Races in the Struggle for Life. New York: The New American Library of World Literature, 1958. (Originally published 1859).

Dickerson, G. E.: Genetic slippage in response to selection for multiple objectives. Cold Spring Harbor Symp. Quant. Biol. 20, 213-224 (1955).

Dickerson, G. E.: Techniques for research in quantitative animal genetics. In: Techniques and Procedures in Animal Production Research. American Society of Animal Production, 1959, pp. 56-105.

Dingle, H.: Life history and population consequences of density, photoperiod, and temperature in a migrant insect, the milkweed bug *Oncopeltus*. Am. Nat. 103, 149-163 (1968).

Dingle, H.: Migration and diapause in tropical, temperate, and island milkweed bugs. In: Evolution of Insect Migration and Diapause. Dingle, H. (ed.). New York: Springer-Verlag, 1978, pp. 254-276.

Dingle, H., Alden, B. M., Blakley, N. R., Kopec, D., Miller, E. R.: Variation in photoperiodic response within and among species of milkweed bugs (*Oncopeltus*). Evolution 34, 356-370 (1980 a).

Dingle, H., Blakley, N. R., Miller, E. R.: Variation in body size and flight performance in milkweed bugs (*Oncopeltus*). Evolution 34, 371-385 (1980 b).

Dingle, H., Brown, C. K., Hegmann, J. P.: The nature of genetic variance influencing photoperiodic diapause in a migrant insect, *Oncopeltus fasciatus*. Am. Nat. 111, 1047-1059 (1977).

Falconer, D. S.: Introduction to Quantitative Genetics. Edinburgh: Oliver & Boyd, 1960.

Finn, J. P.: A General Model for Multivariate Analysis. New York: Holt, Rinehart and Winston, 1974.

Fisher, R. A.: The Genetical Theory of Natural Selection. New York: Dover, 1958.

Hairston, N. G., Tinkle, D. W., Wilbur, H. M.: Natural selection and the parameters of population growth. J. Wildlife Mgmt. 34, 681-690 (1970).

Hegmann, J. P., Possidente, B.: Estimating genetic correlations from inbred stains. Behav. Genet. 11, 103-113 (1981).

Istock, C. A.: Fitness variation in a natural population. In: Evolution of Insect Migration and Diapause. Dingle, H. (ed.). New York: Springer-Verlag, 1978, pp. 171-190.

Lande, R.: Quantitative genetic analysis of multivariate evolution, applied to brain: body size allometry. Evolution 33, 402-416 (1979).

Lande, R.: Genetic variation and phenotypic evolution during allopatric speciation. Am. Nat. 116, 463-479 (1980).

McLaren, I. A.: Inheritance of demographic and production parameters in the marine copepod *Eurytemora herdmani*. Biol. Bull. 151, 200-213 (1976).

Osborne, R., Patterson, W. S. B.: On the sampling variance of heritability estimates derived from variance analysis. Proc. Roy. Soc. Edinburgh, B Ser. 64, 456-461 (1952).

Robertson, A.: Selection in animals: Synthesis. Cold Spring Harbor Symp. Quant. Biol. 20, 225-229 (1955).

Roberts, R. C.: Some evolutionary implications of behavior. Can. J. Genet. Cytol. 9: 419-435 (1967).

PART FIVE

Life History Variation Among Populations

Variation in life histories occurs not only within populations of the same species, but also among geographical races. Local adaptation to prevailing environments has been noted often; a good example is discussed comprehensively by the Taubers in Chapter 4, in Part Two of this volume. The available evidence from studies of geographical variation in life history traits suggests major polygenic influences. The two chapters in this section explicitly examine the genetic structure of life history components in populations of different geographical origins.

Giesel, Murphy, and Manlove consider the life history "strategies" of three populations of *Drosophila melanogaster* reared at different temperatures. Strategies do differ but are a function of temperature because of variation in genetically encoded thermal relations among populations. There is also evidence of genetic trade-offs involving various combinations of life history characters. High heritabilities of some fitness traits may be maintained by a kind of balancing selection resulting from genetic correlations. Using population crosses between two geographical populations of the milkweed bugs, *Oncopeltus fasciatus* and *Lygaeus kalmii*, Dingle, Blau, Brown, and Hegmann examine the genetic structure underlying the life histories. They demonstrate that traits vary in the modes of genetic differentiation and, along with Giesel et al., that the kind of expression may be characteristic of particular environments. They emphasize the dynamic nature of genetic structure, and the way it influences coadaptations which may vary in importance during the process of population differentiation.

Chapter 11

An Investigation of the Effects of Temperature on the Genetic Organization of Life History Indices in Three Populations of *Drosophila melanogaster*

JAMES T. GIESEL, PATRICIA MURPHY, and MICHAEL MANLOVE*

Introduction

Since its conception in a paper by Lamont Cole some 25 years ago, life history theory has developed into a complex field with many hypotheses, approaches, models, and assumptions. The basic question embodied in this set of theory concerns the evolution of life history schedules, including the expected relationships between early life fecundity and late life fecundity or age of reproductive senescence and death. The predicted relationships between these parameters may well depend on the theoreticians' assumptions about the genetic correlation structure of life history traits. For instance, Lewontin (1965) asked how colonizing species might best maximize their intrinsic rate of increase and concluded that they should evolve to reproduce heavily early in life. His analysis was based on the implicit assumption of little or no genetic correlation between life history traits since each trait was considered separately. Murphy (1968) suggested that the distribution of reproductive output over several age classes might be advantageous in situations in which the probability of successful reproduction was temporally variable and unpredictable. Mertz (1971) suggested that the variance in reproductive output about the age of peak reproduction should match environmental uncertainty for successful reproduction. Demetrius (1975) expanded this idea to include matching the entropy of the fecundity schedule to the entropy of the environment. These approaches seem to assume positive genetic correlation between early and late life fitness traits and fecundities.

All of these approaches to the problem are relatively free of any assumptions concerning the cost of reproduction or reproductive effort in terms of reduction in the probability of future reproductive ability. However, Gadgil and Bossert (1970) suggested that genotypes that reproduced heavily early in life might have to pay for this in terms of reduced longevity or subsequent ability to reproduce. Their paper introduced the concept of reproductive effort to the life history theoretician. Heavy reproductive effort early in life was assumed to entail some cost that prevented concurrent maximization of late life fecundity or of longevity. This trade-off assumption, which

*Department of Zoology, University of Florida, Gainesville, Florida 32611 U.S.A.

implies a negative genetic correlation between early and late life fitness traits, has become an entrenched paradigm.

Schaffer (1974), assuming the validity of this concept, introduced the idea that life histories might be molded by the relative extents to which mortality applied to pre-reproductive individuals vs. those which were already reproductively competent, concluding that when juveniles suffered unpredictably high death rates natural selection would lead to the evolution of many ages of reproduction. Conversely, in those populations in which mortality falls most heavily on reproductive adults, there should be selection for early reproduction at the expense of future reproductive potential. Earlier, Williams (1957), advancing one of the many theories of the evolution of senescence, had hypothesized that natural selection would favor the accumulation of genes that enhanced early life fitness (survivorship and fecundity) and that these might be imagined to have negative serial pleiotropic effects on longevity, resulting in cessation of life soon after the end of reproductive life.

Dobzhansky (1958) suggested that, like the construction of a cheap watch, selection should result in the accumulation of sufficient fitness early in life to guarantee reproduction sufficient for the persistence of the population and individual's genotype, with senescence amounting to gradual failure following the guarantee period.

Most existing evidence purporting to test the assumptions and conclusions of life history theory is based on phenotypic correlations between characters. For example, Snell and King (1977) reported a negative phenotypic correlation between fecundity and longevity in *Asplanchnia brightwelli*, a rotifer, but this could have resulted from intragenotypic or environmental effects. Many studies, such as those by Tinkle (1969), Tinkle and Ballinger (1972), Tinkle et al. (1970) and Murphy (1968), have involved measurement of pertinent variables in the field and may simply reflect environmental effects on life history organization. Hickman (1974) demonstrated the strength of such environmental effects when he showed that several populations of an alpine plant that differed in terms of apparent response to r and K selection in situ were identical when grown in the greenhouse. As Stearns (1977) notes, the papers of Solbrig and Simpson (1974, 1977) stand alone in indicating real genotypic interpopulational differences in life history organization. They refer only to differences in growth form and devotion of energy to seed production. Mertz (1975) showed that high early reproduction could be selected for in *Tribolium* but was unable to demonstrate correlated reduction in either survivorship or late life reproduction in the selected lines.

In *Drosophila*, positive genetic correlations between life history components have been shown in several studies. Temin (1966) found positive genetic correlation between viability and fertility, and Mukai and Yamazaki (1971) found positive correlation between developmental rate and viability in studies based on hundreds of lines of *Drosophila melanogaster*. In two sets of data Giesel (1979) and Giesel and Zettler (1980) have found positive genetic correlation between early and late life fitness traits to be the rule, using flies raised at $25°C$ which were derived from a natural population from the Palm Beach, Florida area. Yet, Hiraizumi (1960) found that developmental rates and female fertilities in more fit lines were negatively correlated in this species; and Simmons et al. (1980) presented data suggesting that negative pleiotropy might exist between viability and a measure of competitive ability, using flies taken from a cage that had been seeded from a collection made several years previously in Madison, Wisconsin. The differences among these various results may be caused either by the selective history of the experimental animals or by slight differences in culture conditions applied in the various laboratories. The importance of this latter source of differentiation was clearly shown in a study recently completed by Giesel, Murphy, and Manlove (1982) who showed that heritabilities of, and genetic correlations among, a variety of fitness indices, such as early fecundity, late fecundity, and age of death, were strongly dependent upon culture temperature. In some cases genetic correlations of different sign were obtained with different temperature, using sibling flies, and these results suggested important genotype by environment interactions in the determination of phenotype. Clearly, fitness organizations and assessments of reproductive strategy depend, in this species, on the genesis of animals used in a study and on culture conditions. Existing data do not allow any conclusions beyond this eminently unsatisfactory one. Further, Buzzati-Traverso (1955) demonstrated that a variety of genetic associations among life history traits could be obtained from replicate cultures allowed to evolve under similar conditions, suggesting that ample variation in genetic programming existed within a population to allow for considerable evolution of life history organization and that a variety of evolutionary courses might be equivalent, at least under laboratory conditions. The literature has no information on the question of whether different populations, having evolved under presumptively different selective regimes, may have evolved different organizations of life history traits and whether such different evolution, having occurred may be consistent with the current theoretical understanding of the process.

It is within this framework that our study is cast. We wish to investigate whether genetic organizations of life history differ among three populations of *Drosophila melanogaster* chosen for study because their environments, and natural periodicities of occurrence in their environments, suggest that they may have evolved different reproductive strategies. Our analyses were run at three reasonable temperatures because of our hypotheses that the different populations, coming from the far south and far north of the continental United States, might have evolved different temperature optima and that their genetic organizations of life history traits might have been keyed on these.

Materials and Methods

Flies used in this study were third generation (F = 0.5) descendants (isofemale lines) of individuals caught from wild populations. Three climatic-geographical areas were sampled. Of these, the Miami-Palm Beach, Florida sample was the most temporally diverse. Samples were taken in November 1979, January 1980, and March 1980 by Miss Lourdes Rodrigues from fruit trees located within the Miami metropolitan area. Two samples were used from the Melrose, Florida area. Melrose is about 20 miles northeast of Gainesville. The first Melrose sample was collected from a compost pile in the author's back yard in March 1980. The second was collected from a small orange grove located about 1 mile from the first and within the town of Melrose. In both Florida trapping areas attempts were made to collect *Drosophila melanogaster* on a monthly basis from August 1979 through July 1980. Without going into detail, our trapping records suggest that *D. melanogaster* blooms in late fall and again in early spring in the Miami area, each bloom period lasting for about 2 months. In contrast, records indicate that the *D. melanogaster* bloom lasts continuously from early March through November in the Gainesville-Melrose, Florida area.

The Michigan sample was taken in July 1980 from a bait placed on the banks of the Manistee River near where it crosses Michigan Route 20, and about 8 miles west of Grayling. The bait consisted of a lug of local cherries. This sample may be assumed to consist of only local, native flies since baits took several days to attract flies, apparently from the surrounding woods, and no imported fruits were likely to have been deposited in the area, which was removed from any population centers. It has been reported (Throckmorton and H. Band, personal communication) that *D. melanogaster* is active in northern Michigan for only a short period each year beginning in mid-August and lasting through mid-Octover. Our large July collection casts some doubt on this finding. Nevertheless, the bloom period for this population must be relatively short.

It may be concluded that the Palm Beach and northern Michigan populations are active only over short terms each year, whereas the mid-Florida population occupies a broader temporal niche. The former two populations might be expected to have evolved more colonizing types of life history organizations than those from the longer term suitable environment. At the same time, weather records suggest that during the *D. melanogaster* bloom seasons the Palm Beach and Grayling areas have relatively constant and predictable temperature and humidity regimes. The long bloom season in the Gainesville area, in contrast, must certainly be accompanied by wide and generally unpredictable fluctuations in both rainfall and temperature, especially during the spring and fall months when weather fronts pass through the area on an almost weekly basis but conditions are otherwise apparently suitable for persistence of the species. The areas chosen for sampling of genotypes therefore were picked because they promised to maximize the probability of divergent life history evolution. This sampling from different climatic zones did seem to pose a potential experimental problem, however, for we wondered whether the populations might have evolved different thermal preferenda and optima. In light of this worry we chose to collect our data on life history organizations at three ecologically reasonable temperatures. These were 22°C, 25°C, and 28°C.

Each set of observations was initiated by pair mating one female per strain with either a sibling male or a male from an unrelated strain. These paired flies were placed in half-pint milk bottles, where they were allowed to oviposit for 24-hr periods on a large drop of a standard *Drosophila* medium. After an additional 24-hr period groups of larvae judged to be 1-4 hr old were transferred to plastic vials containing about 5 cm^3 of medium. These were incubated at 22°C, 25°C, or 28°C until eclosion, when adult flies were collected twice a day, mated, and placed in new food cups, which were supplemented with a drop of Fleischmann's yeast suspension. Pairs were transferred to new medium once a day at about noon when eggs were counted. Transferring and egg counting were continued on a daily basis until the females died. New males were added as original ones died in order to assure that experimental females always had mates. In this way complete records consisting of developmental rate, daily fecundity, and age of death of some 200-300 females were collected for each temperature treatment and population.

The data set eventually consisted of records of life history for about 40 strains per population and temperature treatment with an average of four or five replicate female records per strain and treatment. About one-half of the strains were 50% inbred and half were F_1 heterozygotes with inbreeding coefficients of zero. In the absence of a priori reason to expect differences in correlation arising from level of inbreeding, the data were pooled for analysis. Because statistical treatment of this data set would have been unnecessarily cumbersome, each record was reduced to provide the following indices of life history and reproductive performance: developmental rate (DR) = 1/development time, reproduction of day 2 (RD2), age of peak reproduction (APR), reproduction on the day of peak reproduction (PR), and last third of life average daily fecundity (LTF) and age of death (AOD). This choice of indices was obviously a compromise and rather arbitrary.

The reduced records were then analyzed with the aid of a least squares, maximum likelihood analysis of covariance computer program written by Harvey (1977) which estimated heritabilities of individual variables and the genetic (among lines), phenotypic, and "environmental" (within lines) correlations of pairs of variables (Table 11-1). Our intents are to define, within the limits of sample size, the genetic associations of life history parameters of the three populations, to contrast them, and to investigate the extents to which temperature, defining special environmental effects, affects the expression of the genotype. This latter goal of analysis is only possible because sibling flies have been raised under the three environmental regimes. In interpreting the analysis of temperature effect we and the reader must take into account the fact that the lines were not 100% inbred; "replicates" within lines probably in fact represent slightly different genotypes.

Results

Phenotypic organizations of life history variables are shown for the three populations in Figure 11-1A-C. The populations differ significantly in several aspects of life history phenotype, but the differences are dependent on temperature. For example, the Palm Beach and Melrose populations have the same age of peak reproduction and developmental rate at $22°C$, but the latter flies have lower middle and late life fecundity and greater longevity than the flies from Palm Beach. The Grayling flies have a slightly later age of peak reproduction than the two southern populations and an intermediate age of death. At $25°C$ the representatives of the Melrose and Grayling populations have higher early reproduction than the representatives of the Palm Beach population. They also have earlier ages of peak reproduction and shorter life spans. The Grayling flies have about 30% higher peak fecundity than either southern population. At $28°C$ the average Palm Beach fly has an earlier age of peak reproduction than flies from either more northern population. The Grayling flies have high peak fecundity. Flies from both of these early reproducing populations have shorter life spans than the Melrose flies.

By rearranging the data as in Figure 11-2A-C the effects of the three temperatures on the life history phenotypes and genetic correlations of the life history parameters can be examined more fully. In Figure 11-2A, the life history phenotypes of the Palm Beach sample are found to be quite similar at $22°C$ and $25°C$, both sets of flies reproducing relatively late in life, but the $28°C$ regime results in striking modifications in life history. At $28°C$, the average fly shows a classic "colonizing" type of life history distribution. Accompanying this shift in average performance no significant shifts in

Table 11-1. Heritabilities of life history indices estimated for the three populations[a]

Index	22°C			25°C			28°C		
	Melrose	Palm Beach	Grayling	Melrose	Palm Beach	Grayling	Melrose	Palm Beach	Grayling
DR	0.993	1.0	1.0	0.708	1.0	1.0	0.386	0.503	0.801
RD2	0.053	1.0	0.496	0.772	0.433	1.0	0.465	0.870	0.609
PR	0.412	0.495	0.471	1.0	0.497	0.764	0.794	0.240	0.857
APR	0.519	0.287	0.680	0.155 n.s.	0.821	0.591	0.145 n.s.	0.186	0.708
LTF	0.831	0.284	0.383	0.360	0.413	0.219 n.s.	0.365	0.293	0.219 n.s.
AOD	0.318	0.264	0.474	0.226	0.004 n.s.	0.208 n.s.	0.617	0.299	0.173 n.s.

[a]Flies at each temperature are full siblings of those at the other temperatures. DR, developmental rate; RD2, reproduction on day 2; PR, peak fecundity; APR, age of peak reproduction; LTF, last third of life average daily fecundity; AOD, age of death. n.s., not significant.

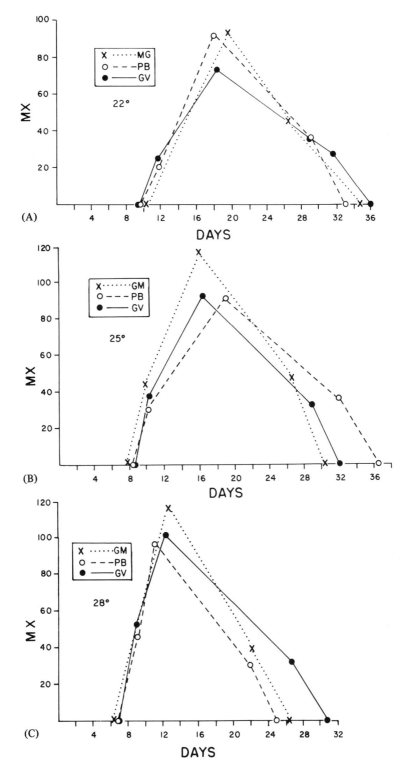

Figure 11-1. Average net fecundity ($l_x m_x$) distributions for the three populations of flies. A, $22°C$; B, $25°C$; C, $28°C$.

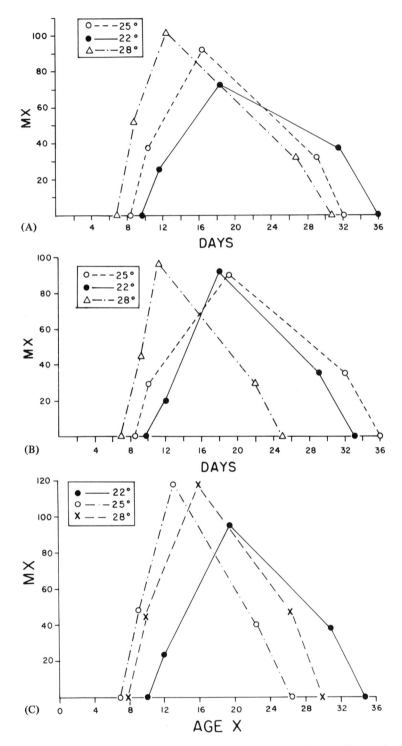

Figure 11-2. Net fecundity (m_x) distributions arranged to show effects of rearing temperatures on each sample. A, Melrose; B, Palm Beach; C, Grayling.

genetic association are found between the quantity variables (peak reproduction, reproduction on day 2, and last third of life average per day fecundity) (Table 11-2). In contrast there seem to be clear trends in the relationships of quantity variables with timing variables. There is a consistent shift from a positive genetic correlation (0.204) between RD2 and APR at 22°C (Table 11-2) to negative genetic correlation at 25°C (-0.258) and 28°C (-0.754). Similarly, APR is negatively correlated with peak reproduction and last third fecundity at 22°C (-0.429) and these associations become more positive as rearing temperature is increased. Age of peak reproduction is positively correlated with age of death at 22°C (0.697) but the association is negative at 28°C (-0.497). This difference in genetic correlation is statistically significant, as is the shift in genetic correlation between peak reproduction and age of death. The genetic program, common to the flies at all three temperatures, is clearly expressed in different ways under these special environmental conditions. Speculation as to the theoretical reasons for these differences in expression of the genotype would be premature at this point.

The Melrose data suggest an obvious difference in the effects of temperature on the life history phenotype of the average fly (Fig. 11-2A). In the Palm Beach material there was an abrupt shift in phenotype with a break occurring in the 28-25°C range. No such abrupt shift is apparent in the Melrose material. Instead, there is a steady progression from high early reproduction coupled with a relatively short life span at 28°C to a lower, delayed peak of reproductive activity at 22°C.

The genetic correlation between development rate and age of death for Melrose is significantly positive at 22°C but significantly negative at 28°C. Apparently, flies that are genetically programmed to develop quickly are more longevous when raised at low temperatures but die earlier at the higher temperature. Associations of peak fecundity and age of peak fecundity with late fecundity and age of death are positive regardless of temperature. Note, however, that positive correlation of age of peak fecundity with other variables signals a trade off in fitness space; genotypes that have an early age of peak fecundity tend to die earlier, etc.

The shift from 22°C to 28°C growing temperature also causes a shift from positive correlation of early reproduction with late fecundity to negative correlation. It is easy to suggest that this reflects greater stress attendant on early reproduction at the higher temperature. These data illustrate genotype by environment interaction in the determination of phenotype, a phenomenon that has been investigated seldom but is important to those of us who are concerned with determining whether populations differ in the genetic organization of life history traits or in reproductive strategies.

The Grayling data offer further basis for comparison of temperature effects. The shift from 28°C to 25°C rearing temperature results in increased age of peak reproduction and longevity but no decline in age-specific fecundity levels, seeming to result simply in a change in timing of reproductive activity at all ages which is accompanied by delayed age of peak reproduction and greatly increased life span. The effects of temperature on correlations characteristic of the Michigan population are quite different from those we noted for either southern population. The quantity variables are all positively correlated with each other at 25°C and 28°C as they were for the southern populations. However, the genetic correlations of reproduction on day 2 with peak reproduction and last third fecundity become strongly negative at 22°C, signaling a genetic trade-off between amounts of early reproduction and late reproduction at this

Table 11-2. Genetic correlations (r_g) between fitness indices as determined for flies raised at three temperatures

Index	Melrose sample 22°C	25°C	28°C	Index	Palm Beach sample 22°C	25°C	28°C	Index	Grayling sample 22°C	25°C	28°C
DR-RD2	0.116	0.463[a]	-0.176	DR-RD2	0.411[a]	0.124	0.331	DR-RD2	-0.020	0.271	0.260
-APR	-0.033	0.429	-0.002	-APR	-0.13	0.153	0.142	-PR	0.031	0.590[a]	0.537[a]
-PR	-0.027	0.169	-0.359	-PR	0.251	0.621[a]	0.268	-APR	0.109	-0.147	-0.090
-LTF	-0.214	-0.238	0.148	-LTF	-0.004	0.486[a]	0.251	-LTF	-0.154	-0.402	0.417
-AOD	0.763[a]	0.647	-0.474[a]	-AOD	0.126		0.583	-AOD	0.002	0.882[a]	1.0[a]
RD2-APR	-0.800	0.270	-0.876	RD2-APR	0.204	-0.258	-0.754[a]	RD2-APR	-0.518[a]	-0.072	-0.379
-PR	0.539	0.592[a]	0.453[a]	-PR	0.332	0.619[a]	0.373	-PR	-0.358	0.318	0.670[a]
-LTF	0.465	0.221	-0.352	-LTF	0.197	0.545[a]	0.173	-LTF	-0.249	0.734	0.313
-AOD	-1.0	1.0	0.529[a]	-AOD	0.348	0.082	0.766[a]	-AOD	-0.380	0.314	0.973[a]
APR-PR	0.272	0.823[a]	0.264	APR-PR	-0.314	-0.039	0.087	APR-PR	0.352	0.097	-0.103
-LTF	0.402[a]	1.0[a]	0.564[a]	-LTF	-0.429	0.000	0.220	-LTF	-0.080	0.089	-0.008
-AOD	0.687	1.0	0.023	-AOD	0.697[a]		-0.497	-AOD	0.442[a]	0.468	0.507
PR-LTF	0.881[a]	1.0[a]	0.763[a]	PR-LTF	0.902[a]	1.0[a]	0.624[a]	PR-LTF	0.437	0.236	0.942
-AOD	0.349	1.0	0.697[a]	-AOD	-0.346		0.256	-AOD	0.162	0.812	0.965[a]

[a] Correlations that are significant, at least p = 0.05.

temperature. The shift in the genetic relationship between RD2 and PR seems to be part of a steady progression. Amount of early reproduction is positively correlated with longevity at 25°C and 28°C but those strains which reproduce heavily early in life at 22°C tend to die earlier than those which have lower early fecundity. Similarly, strains that develop rapidly are more longevous at 25°C and 28°C but there is no apparent relationship between developmental rate and longevity at 22°C. At 28°C rapidly developing strains reproduce heavily late in life but have lower late reproduction than strains that develop more slowly at 25°C and 22°C. At the same time, rapidly developing strains show a slight tendency toward early age of peak reproduction at 25°C and 28°C but have later age of peak reproduction than slower developing strains at 22°C. Age of peak reproduction is negatively correlated with reproduction on day 2 at all temperatures. This suggests that the Michigan flies, in contrast to the southern strains, are programmed such that those which reproduce heavily very early in life also have an early age of peak reproduction. Strains that have an early peak of reproductive activity are less longevous than those that are genetically programmed to peak reproductively at a later age.

The estimates of genetic correlation presented in Table 11-2 suggest that the three populations may have different kinds of genotype by environment interactions and that these may be based on different critical temperatures. The populations appear to differ in the effects of temperature on the relationships between:

(1) developmental rate and age of death. In the Palm Beach and Grayling material these variables become more positively correlated with increases in temperature; the Melrose material shows positive genetic correlation between these variables at 22°C with development of negative genetic correlation at 28°C.

(2) reproduction on day 2 and age of peak reproduction. The association between these variables is negative at all temperatures for Michigan flies but is more strongly negative at 28°C and slightly positive at 22°C in the Palm Beach material and weakly positive at 25°C but negative at both 22°C and 28°C in the Melrose sample.

(3) reproduction on day 2 and last third of life average per day fecundity. These variables show a trend from positive genetic correlation at low temperature to negative genetic correlation at 28°C in the Melrose genotypes, are negatively correlated at low temperature but positively correlated at high temperature in the Michigan material, and are positively correlated at all three temperatures in the Palm Beach material.

(4) age of peak reproduction and age of death. The genetic correlation between these variables is positive at low temperature but negative at high temperature in Palm Beach flies but positive at all temperatures in Michigan and Melrose strains.

(5) reproduction on day 2 and age of death. These are positively correlated at all temperatures in the Palm Beach material but negatively correlated at 22°C in both the Michigan and Melrose material.

The Palm Beach and Grayling populations show more tendency toward negative genetic correlation of variables at 22°C but positive genetic correlation at higher temperatures; the Melrose population tends to exhibit positive genetic correlations between variables at 25°C but negative correlations at the extreme temperatures. These populations therefore may have different thermal relations (optimal ranges) which are reflected in their different modes of temperature dependency of the genetic correlations.

We cannot close our presentation of results without drawing attention to the full extent of the genotype by environment interactions in the determination of phenotype. The changes in genetic correlation structure of fitness components attendant on differences in rearing temperature are one important aspect of this. They show that apparent reproductive strategy is dependent on culture conditions and make one wonder whether, indeed, any of the populations can be said to have a consistent reproductive strategy. Certainly the populations can be said to indulge in bet hedging, having the capacity to modify life history form in ways that may be consistent with environmental conditions. Whether this results from different expressions of the same components of the genome at the different temperatures or, perhaps, from expressions of different segments of the genome cued to act by the different temperatures awaits future investigation.

Nevertheless, the extent of environmental effects on the expression of individual genotypes is another important aspect of the system. In Figure 11-3 we show the phenotypes of several different Palm Beach genotypes, as defined by values of age of peak reproduction and last third fecundity at $22°C$ and $28°C$. It may be recalled that, overall, the genetic correlations between these two parameters were positive at $22°C$ and negative at $28°C$. Several lines, such as 4, 6, and 17, which are characterized by low late fecundity at $22°C$, show substantial increases in this parameter at $28°C$. This effect of temperature on late fecundity seems to be related to $22°C$ performance, for lines that have intermediate values of both parameters show weaker responses to the temperature shift. It might be concluded that the lines which reproduced more heavily at $22°C$ were suffering at $28°C$ because of the stress induced by the high temperature and attendant increase in rate of metabolism and that only evidence of physiological trade-off had been found. Nevertheless, the effects of the temperature shift are not totally consistent or predictable; such lines as 16, 1, and 13 show effects of temperature that are not predictable from knowledge of the more common relationship.

Perhaps an underlying phenomenon awaits discovery here, but for now we draw attention to the lack of predictability shown. The Melrose genotypes, although offering about the same predictability of $22°C$ performance on these axes for $28°C$ performance as do the Palm Beach data, possibly suggest a different set of temperature effects. In the Melrose data it seems that genotypes that have low values of age of peak reproduction and last third fecundity at $22°C$ tend to have even lower values of late fecundity at $28°C$. In contrast, there may be a general pattern of increase in late fecundity among genotypes that have high values at $22°C$. Again, a large fraction of the lines fail to fit this general pattern and evolution, acting on either parameter of fitness at either temperature, may be expected to produce the opposite results at the other temperature.

In order to investigate the forms of these temperature effects on genotypic performance further, we chose the sets of genotypes circled in Figures 11-3 and 11-4 for computation of complete average m_X distributions, treating each as a group rather than using individual genotypes. The resultant m_X distributions are shown in Figure 11-5 where their genesis from Figures 11-3 and 11-4 is shown in the insets. Those genotypes which have early peaks of reproduction and are short lived at $22°C$ show increases in the values of all parameters when raised at $28°C$, whereas those which

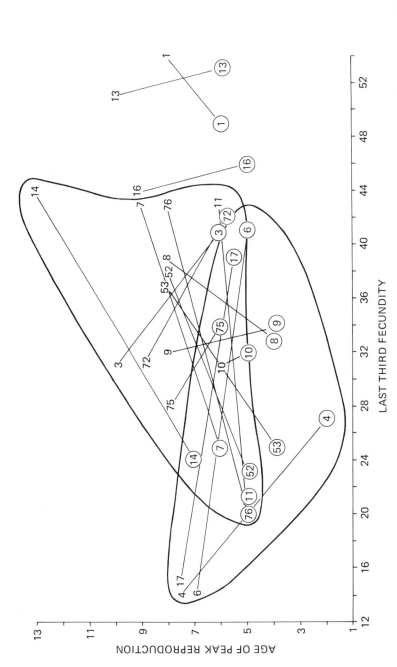

Figure 11-3. Effects of temperature on the relationships between age of peak reproduction and last third fecundity for genotypes of the Palm Beach sample; uncircled numbers, 22°C phenotypes; circled numbers, 28°C phenotypes. Lines connect 22°C and 28°C values for each strain. The envelopes designate sets of strains that seem to show the same form of effects of environment of genotypic expression (i.e., the same mode of genotype by environment interaction effect).

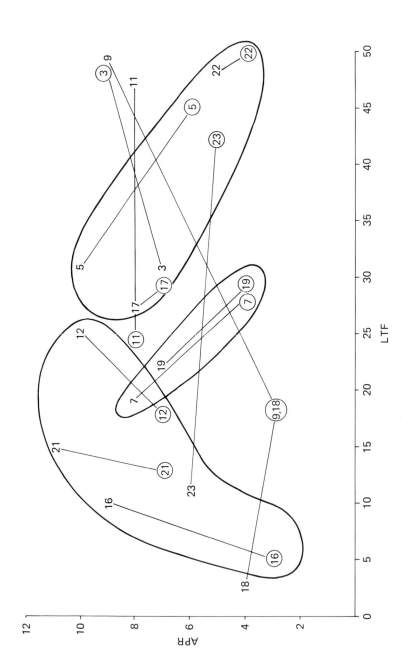

Figure 11-4. Effects of temperature on the relationships between age of peak reproduction and last third fecundity for the Melrose sample. Labeling as in Figure 11-3.

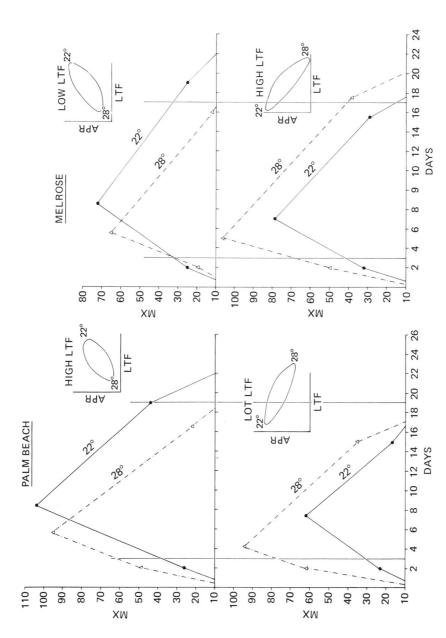

Figure 11-5. Net fecundity distributions of circled sets of strains from Figures 11-3 and 11-4. Lines drawn at age 3 represent selection thresholds. Consideration of these is useful in investigating the effects of the genotype-environment interaction on probably correlated effects of selection on early fecundity level.

reproduce later at 22°C show decreased age of peak reproduction and increased early fecundity, which occurs at the expense of late fecundity and longevity. Perhaps this signals a polymorphism for thermal optimum that occurs in both populations. Perhaps it signals some deeper seated common physiological relationship (based on rate of metabolism?). Interestingly, although it is possible to discern a relationship between timing of life history events at 22°C and the effects of the 28°C regime on the genotype-phenotype relationship, there does not seem to be any association between quantity of fecundity at 22°C and the physiological effect of the temperature shift.

In any event these graphs, representing the extreme genotypes in the two populations, give us a good chance to illustrate the point that the two populations may be expected to respond to selection on life history traits in different ways, at different temperatures. In Figure 11-5 we draw selection thresholds at day 3 of adult life. To investigate the probable results of selection for high early reproduction (being able to reproduce early in life), it is only necessary to count the numbers of eggs the classes of genotype can be expected to produce during the first 3 days of life, and then to calculate relative reproductive rates for the two classes of genotype in a population. This will give a crude representation of the population one discrete generation hence. It can be seen that when the Palm Beach population is selected for early reproductive ability at 22°C, the result is a larger proportion of the population than before that has a late peak of reproductive activity and is more longevous. Similar selection carried out at 28°C produces a population that has an earlier age of peak reproduction and is shorter lived than prior to selection. In the Melrose population earlier reproduction and shorter life spans are expected to result from selection at 22°C and overall increases in fitness to result from selection at 28°C. The populations differ, according to this analysis, in responses to selection at the same temperatures and in the effects of temperature on predicted response to selection. These predicted responses to selection are essentially identical to ones that can be made using the genetic correlations presented earlier.

Discussion

In our introduction we suggested that the Melrose, Palm Beach, and Grayling populations might be expected to be characterized by different genetic associations between life history traits and to exhibit different "life history strategies." Our predictions were based on knowledge, however incomplete, of the environments characteristic of the areas and of the yearly periodicities of activity of the populations. On this basis we predicted that the Palm Beach and Grayling populations should have evolved colonizing life history distributions. According to trade-off theory these should be reflected by negative genetic correlations between indices of early life fitness (development rate and early fecundity) and late life fitness (last third fecundity, age of death). If the trade-off paradigm is not applicable, a colonizing strategy should be reflected in high values of all parameters or should consist of high early fecundity, which is unassociated genetically with values of late life reproductive traits (assuming that life history traits can evolve independently, as assumed implicitly by several authors). The Melrose population was expected, assuming the validity of the theories of Mertz (1971) and Demetrius (1975), to have more ages of reproduction and to spread reproductive effort more evenly over the life span than the two more colonizing populations.

The data support the predictions to some extent. The Grayling population reproduces more early in life than either of the southern populations and the Palm Beach population, on the average (taken across the three temperatures), reproduces earlier than the Melrose population. However, when the temperature-specific performances of the latter two populations are considered, a complication is noted. The life history phenotypes of these populations are dependent on temperature in different ways, such that a population that may be characterized as having a more colonizing phenotype at one temperature has, relatively speaking, later average fecundity at another. Because these phenotypes define the populations' performances under constant conditions and the populations differ, it may be concluded that the populations do in fact have different life history strategies but that they are dependent on temperature. These differences may result primarily from different genetically encoded thermal relations of the populations.

Consideration of the genetic correlations between fitness indices characteristic of the populations allows us to make several theoretically interesting points. First, we find clear evidence of genetic trade-off of age of peak reproduction with peak reproduction, the last third fecundity, and age of death. In the Melrose material these correlations are positive regardless of temperature. The positive genetic correlations suggest that genotypes that have early peaks of reproductive activity also have low reproduction, last third fecundity, and age of death. It might be concluded that such genotypes "pay for" their reproductive precocity with lessened ability to produce young at the age of peak reproduction and later ages and with decreased life span. In the other two populations the expressions of these trade-offs are temperature dependent. For example, age of death is positively correlated with age of peak reproduction at all temperatures in the Grayling material, but the trade-off applies only at 22°C in the Palm Beach material. These variables are negatively correlated at 28°C in the Palm Beach flies. Similarly peak reproduction and last third fecundity are positively correlated with age of peak reproduction at 28°C in the Palm Beach sample but negatively correlated at 22°C where, apparently, strains that have an early peak of reproductive activity also have high peak fecundity and late fecundity. Genetic correlations of development rate and RD2 (early reproduction) are similarly temperature dependent, with trade-offs being indicated in one environment, positive genetic correlations applying in another, and the effects of temperature being population specific. It is particularly interesting in this light to note that Grayling strains which have high early fecundity are more longevous at 28°C but less longevous at 22°C. The three populations differ, clearly, in the effects of temperature on genetic expression of association of fitness indices.

However, these data go much further than a simple "test" of these basic assumptions of the theory of life history evolution. Throughout the analysis of these data we have been struck by the pervasive influences of special environment upon the expression of the genotype. In terms of basic, mean life history distributions the three populations differ in different ways at different temperatures. At the same time we have found that certain of the genetic correlations between life history indices are labile to temperature. These changes in genetic correlation signify the potential for condition-dependent evolution of the populations in question, with correlated responses to selection being functions of thermal regime. Finally, as in Figure 11-5, we have been able to demonstrate that the phenotypes characteristic of individual genotypes are strongly modifiable by rearing temperature. Because these modifications were not simple responses to temperature but were dependent on the original phenotypes of a genotype, they again demonstrated the importance of genotype by environment interaction in the determination of phenotype. The analysis of predicted correlated responses to selection carried out with the aid of this figure demonstrates the importance of considering genotype by environment interactions when assessing the probable evolutionary history and course of the populations.

Finally, we should like to discuss what these data have to tell about conditions necessary for the maintenance of genetic variability for life history traits. Usually (R. Lande, personal communication) it is assumed that negative genetic correlation or serial pleiotropy (see Williams 1957) is necessary for maintenance of genetic variability, because directional selection for, say, high early reproduction will be countered in the presence of negative pleiotropy by reduction in the values of correlated traits and conditions sufficient to produce a balanced polymorphism may result. If genotypic expression is environment dependent in the ways we have shown it to be, however, we may expect that even when selection per se is directional the fact that different genotypes are selected under different conditions also is likely to result in balanced polymorphism. The operation of this sort of balancing selection may be indicated by the high heritabilities of the fitness traits we have measured.

Investigation of the genetic organization of life history (and the "life history strategies") of populations is not amenable to simple analysis. If application of three temperatures to the genotype can cause such remarkable changes in its expression, we cannot help but wonder what other environmental variables may also act to produce special environmental effects. Given a reasonably complete picture we may then begin to ponder how the life histories of populations evolve to differ, and indeed, given the vicissitudes of natural environments how different we can expect them to become.

References

Buzzati-Traverso, A. A.: Evolutionary changes in components of fitness and other polygenic traits in *Drosophila melanogaster* populations. Heredity 9, 153-186 (1955).

Demetrius, L.: Reproductive strategies and natural selection. Am. Nat. 109, 243-249 (1975).

Dobzhansky, T.: Genetics of homeostasis and senility. Ann. N.Y. Acad. Sci. 71, 1234-1241 (1958).

Gadgil, M., Bossert, W.: Life historical consequences of natural selection. Am. Nat. 104, 1-24 (1970).

Giesel, J. T.: Genetic co-variational survivorship and other fitness indices in *Drosophila melanogaster*. Exptl. Gerontol. 14, 323-328 (1979).

Giesel, J. T., Zettler, E.: Genetic correlations of life historical parameters and certain fitness indices in *D. melanogaster*: r_m, r_s, and diet breadth. Oecologia 47, 299-302 (1980).

Giesel, J. T., Murphy, P. H., Manlove, M. The influence of temperature on genetic interrelationships of life history traits in a population of *Drosophila melanogaster*: What tangled data sets we weave. Am. Nat. 119, 464-469 (1982).

Harvey, W. R.: Users Guide for LSML 76. Mixed model least square and maximum likelihood computer program. Available from Walter Harvey, Department of Dairy Science, Ohio State University, Columbus, Ohio, 1977.

Hickman, J. C.: Environmental unpredictability and the plastic energy allocation strategies in the annual *Polygonum cascadense* (Polygonaceae). J. Ecol. 63, 689-702 (1974).

Hiraizumi, Y.: Negative correlation between rate of development and female fertility in *Drosophila melanogaster*. Genetics 46, 615-624 (1960).

Lewontin, R. C.: Selection for colonizing ability. In: The Genetics of Colonizing Species. Baker, H. G., Stebbins, G. L. (eds.). New York: Academic Press, 1965, pp. 77-94.

Mertz, D. B.: Life history phenomena in increasing and decreasing populations in statistical ecology. In: Sampling and Modeling Biological Populations and Population Dynamics. Vol. 2. Patil, G. P., Pielou, E. C., Waters, W. E. (eds.). University Park, Pa.: Pennsylvania State University Press, 1971, pp. 361-399.

Mertz, D. B.: Senescent decline in flour beetle strains selected for early adult fitness. Physiol. Zool. 49, 1-23 (1975).

Mukai, T., Yamazaki, T.: The genetic structure of natural populations of *Drosophila melanogaster*. X. Development time and viability. Genetics 69, 385-398 (1971).

Murphy, G. I.: Pattern in life history and the environment. Am. Nat. 102, 391-403 (1968).

Schaffer, W. M.: Selection for optimal life histories: The effects of age structure. Ecology 55, 291-303 (1974).

Simmons, M. J., Preston, C. R., Engles, W. R.: Pleiotropic effects on fitness of mutations affecting viability in *Drosophila melanogaster*. Genetics 94, 467-475 (1980).

Snell, T. W., King, C. E.: Lifespan and fecundity patterns in rotifers: The cost of reproduction. Evolution 31, 882-890 (1977).

Solbrig, O. T., Simpson, B. B.: Components of regulation of a population of dandelions in Michigan. J. Ecol. 63, 473-486 (1974).

Solbrig, O. T., Simpson, B. B.: A garden experiment on competition between biotypes of the common dandelion *Taraxicum officinale*. J. Ecol. 65, 427-430 (1977).

Stearns, S. C.: The evolution of life history traits: A critique of the theory and a review of the data. Ann. Rev. Ecol. Syst. 8, 145-171 (1977).

Temin, R. G.: Homozygous viability and fertility loads in *Drosophila melanogaster*. Genetics 53, 27-46 (1966).

Tinkle, D. W.: The concept of reproductive effort and its relation to the evolution of life histories of lizards. Am. Nat. 103, 501-516 (1969).

Tinkle, D. W., Ballinger, R. E.: *Sceloporus undulatus*: A study of the intraspecific comparative demography of a lizard. Ecology 53, 570-585 (1972).

Tinkle, D. W., Wilbur, H. M., Tilley, S. G.: Evolutionary strategies in lizard reproduction. Evolution 24, 55-74 (1970).

Williams, G. C.: Pleiotropy, natural selection, and the evolution of senescence. Evolution 11, 398-411 (1957).

Chapter 12

Population Crosses and the Genetic Structure of Milkweed Bug Life Histories

HUGH DINGLE, WILLIAM S. BLAU, CARL KICE BROWN, and
JOSEPH P. HEGMANN*

Introduction

The fitness of a given phenotype is a direct consequence of the schedule of behavior, fecundity, and mortality that constitutes a life history. For this reason life histories are major adaptations of unique importance to general Darwinism (Bell 1980). As with other complex adaptations life histories consist not of single characters, but of sets of phenotypic traits that covary and function together (Frazetta 1975). Such sets often are referred to as "strategies" or "tactics" and much theoretical and empirical effort has been devoted to attempting to understand the evolution of the complex known as a "life history strategy" (Bell 1980, Stearns 1976, 1977). Births and deaths are most closely related to fitness and have drawn most of the attention, but behavior is also an important component of life histories, especially as it confers flexibility on where and when to breed (Istock 1978 and Chapter 1, this volume, Nichols et al. 1976, Taylor and Taylor 1977, 1978). Two important elements of insect behavior are migration and diapause (Dingle 1981, Solbreck 1978), and we consider them in our discussion here, along with the more traditional life table statistics that they influence.

A genetic analysis is central to the complete understanding of any adaptation because genetic variance provides the raw material upon which natural selection can act. The subject of this and our previous chapter (Hegmann and Dingle, Chapter 10, this volume) is a quantitative analysis of the relation between genetic structure and phenotypic covariances, i.e., how genes influence how life history traits function together. We examine this question in crosses between widely separated populations of the milkweed bugs *Oncopeltus fasciatus* and *Lygaeus kalmii*. We compare an assortment of life history characters measured on parents and their "hybrid" and "purebred" offspring and in *L. kalmii* grandoffspring.

The analysis was based on the assumption that each population would be characterized by its own coadapted gene pool. This coadaptation should affect the covariance structure among the array of traits constituting a life history. Much covariation arises

*Program in Evolutionary Ecology and Behavior, Department of Zoology, University of Iowa, Iowa City, Iowa 52242

because genes may interact, and selection is unlikely to influence one gene independently of others (Endler 1977, Lewontin 1974, Sokal and Taylor 1976). This, plus the continuous alteration of gene combinations from generation to generation, should result in the integration of local gene pools through the coadaptation of segregating interacting elements (Endler 1977, Wallace and Vetukhiv 1955). The reassortment of genetic elements in offspring from population crosses should disrupt this integration and permit the analysis of complex adaptations at two levels. The first concerns the nature of mechanisms resulting in genetic coadaptations; the second (and our focus) is how these coadaptations are constructed so that they influence covarying phenotypic complexes such as life histories.

Most investigators have emphasized the first concern, i.e., how genetic coadaptation might occur. Wallace and Vetukhiv (1955), for example, suggested that it could be brought about by epistatic interactions between homozygous loci, heterozygosis for chromosomal configurations (reviewed by Dobzhansky, 1970), or the integration of entire gene pools based on selection acting on genes in a heterozygous state. Interpopulation crosses in *Drosophila* displayed heterotic effects for survivorship and fecundity in the first generation but showed reduced expression in the F_2, suggesting that indeed local gene pools were integrated (Vetukhiv 1956, 1957, Wallace 1955). Reduced viability, indicating the breaking part of coadapted gene combinations, has also been noted for population crosses in Lepidoptera (Oliver 1972, 1979), amphibians (Bachmann 1969), and plants (Kruckeberg 1957, Lindsay and Vickery 1967). That subviability of interpopulation hybrids is often a consequence of developmental disruption suggests that regulatory genes are important in coadaptation (Ohno 1969, Oliver 1979). The frequently observed range of expression of hybrid traits from heterosis to parent-offspring resemblance to developmental disruption can be explained by a model of coadaptation that postulates different effects of modifier genes on major gene genotypes (Endler 1977, Chapter 4).

The second aspect of the analysis of complex adaptations, that involving the relationship between the genetic structure of populations and the phenotypic associations of traits, has received much less attention than the mechanism of coadaptation. It is the aspect we address here and in our previous chapter (Hegmann and Dingle, Chapter 10, this volume). We are trying to analyze, with respect to life histories, what Lewontin (1974, p. 14) called the "phenotypic space" consisting of the simultaneous effects of the same gene differences on sets of, rather than single, characters. Such an attempt requires the application of quantitative genetic methods to connect genetic variation with phenotypic variation as Lande (1979 and Chapter 2, this volume), for example, has emphasized.

This and our earlier chapter in this volume are parts of an ongoing effort to understand the adaptive basis of insect life histories. In this study we crossed, first, populations of *Oncopeltus fasciatus* from Iowa and Puerto Rico, chosen both because they were widely separated geographically and because they displayed a number of environmentally influenced life history differences as indicated below. The offspring of crosses were reared in two photoperiods, because photoperiod can be an important environmental variable influencing life history traits in this species (Dingle 1968, 1974, Dingle et al. 1977). Second, we crossed Iowa and Colorado populations of *Lygaeus kalmii* that are sufficiently morphologically distinct to be considered subspecies (Slater and Knop 1969). They also displayed differences in migratory activity (Caldwell 1974).

The expression of traits in *L. kalmii* was analyzed in both first- and second-generation descendants of the original crosses. The geographical differences observed in *O. fasciatus* and *L. kalmii* suggested the evolution of locally coadapted gene complexes. The variation in the expression of life history traits in the hybrid offspring of both species indicates differences in degree of divergence with respect to gene influences (coadaptation) and is potentially of considerable importance in population differentiation and hence evolution.

Natural History of Puerto Rico and Iowa *Oncopeltus fasciatus*

Bugs of the Iowa population of *Oncopeltus fasciatus* are migrants that arrive to breed in late June or early July, producing one and sometimes a partial second generation on fruiting milkweed (*Asclepias*). Bugs reaching adulthood in late summer and early autumn respond to short days and lower temperatures by entering reproductive diapause (Dingle 1968) and emigrating (Dingle 1981). Tethered flight experiments demonstrate that diapause increases the migratory period in the life cycle (Dingle 1978). The bugs are unable to survive the rigors of an Iowa winter, and prereproductive emigration is adaptive as long as any individuals reach available breeding areas to the south (Dingle et al. 1977).

Our evidence indicates considerable additive genetic variance for the critical photoperiod of the diapause response (Dingle et al. 1977). This is consistent with data from an extensive array of other insect species (Danilevskii 1965, Hoy 1978). In our experiments genetic variance was assessed with a selection experiment and by means of offspring-parent regression, both standard methods of estimation (Falconer 1960). In the selection experiment, Iowa bugs were maintained on an LD 12:12 photoperiod at 23°C, and the earliest females to reproduce were selected as the parents of the next generation. By generation 10, there was essentially no diapause in the selected population. From these data and from concurrent offspring-parent regression analysis, we estimated that approximately 70% of the total phenotypic variance in photoperiodically cued age at first reproduction resulted from additive genetic variance (i.e., heritability was about 0.7).

In contrast to Iowa, breeding sites in Puerto Rico are available throughout the year with consequences for both diapause and flight. Because of human activity there is considerable local variation in the availability of fruiting milkweed (*Asclepias curassavica*). The result is that the bugs usually produce only one generation in any given patch and are constantly undertaking local movements (Dingle 1981). Although in the laboratory a few individuals may show some reproductive delay on short days, the majority show no response to photoperiod (Dingle et al. 1980a), suggesting that diapause is not ordinarily adaptive in Puerto Rico. Further consistent with the respective life histories in the field, the greatest amount of flight was displayed by the Iowa bugs (Dingle et al. 1980b). More Iowa than Puerto Rico bugs flew for longer than 30 min in tethered flight tests, whereas more of the Puerto Rico population failed to fly. Comparisons among several insect species, including *O. fasciatus*, indicate that among closely related forms, individuals of migrant species or populations tend to be larger. This was also true here, as Iowa bugs are significantly larger (as measured by wing length) than those from Puerto Rico.

Differences between the two populations are not limited to photoperiodic and flight responses but also extend to a number of traditional life history statistics (Dingle 1981). These are indicated in Figure 12-1, which displays interaction plots of the responses of the various characters to four combinations of day length and temperature. Diapause in Iowa is indicated in the interaction between temperature and photoperiod for age of first reproduction. In LD 11:13 Iowa bugs delay reproducing to almost 70 days posteclosion at 23°C, but the effect of short days is overridden at 27°C. There is only slight indication of any interaction in the Puerto Rico sample. Population-environment interactions are also displayed by the other traits in Figure 12-1. Clutch size is larger in Puerto Rico with fewer clutches and a greater interclutch interval. The latter appears to be more influenced by temperature in the Puerto Rico bugs, whereas clutch number shows a conspicuous effect of photoperiod in the Iowa population. An interaction effect on life span, reflecting the time spent in diapause, also shows up in the Iowa bugs. Finally, the two populations differ conspicuously in total fecundity; Iowa shows a marked influence of photoperiod, whereas temperature is clearly more important for Puerto Rico.

These differences in life history traits suggested that crosses between the two populations might prove interesting with respect to the underlying gene influences. Endler's modifier gene model (1977 Chapter 4) predicts an array of possible outcomes depending on the nature of the modifiers. We predicted that traits with relatively high additive genetic variance, e.g., age at first reproduction (Dingle et al. 1977) or wing length (Hegmann and Dingle, Chapter 10, this volume), in one of the populations would show intermediate expression in the population hybrids, because such variance should indicate the absence of strong directional selection leading to genic coadaptation (Falconer 1960). Such a result would also be consistent with the literature on the genetics of diapause (Danilevskii 1965, Hoy 1978). Traits with relatively low additive genetic variance should have been subjected to strong selection producing coadaptation, with hybrid offspring thus displaying enhanced (overdominance or "heterotic") or reduced expression depending on the type and extent of genotype differentiation (Endler 1977). Because the two experimental populations respond differently to photoperiod, the expression of some traits also can be expected to be environment dependent. Finally, we predicted similar hybrid expression for traits showing high genetic correlations, such as clutch size and wing length (Hegmann and Dingle, Chapter 10, this volume).

Natural History of *Lygaeus kalmii*

Unlike *O. fasciatus*, *L. kalmii* is a resident of the temperate zone, but the two species do overlap in the same milkweed patches in the summer and fall. The latter, however, overwinters in a temperature-induced low-intensity diapause, and we have failed to demonstrate a photoperiodic response in conditions that induce diapause in *O. fasciatus*. In Iowa there is only local migration to locate diapause and feeding sites. A fraction of the Colorado population shows longer duration tethered flight than Iowa bugs but most fly only for brief periods. This bimodal pattern apparently results from adaptation to the often highly dispersed patches of milkweed, which in the Western United States occurs primarily along rivers or in mountain valleys (Caldwell 1974).

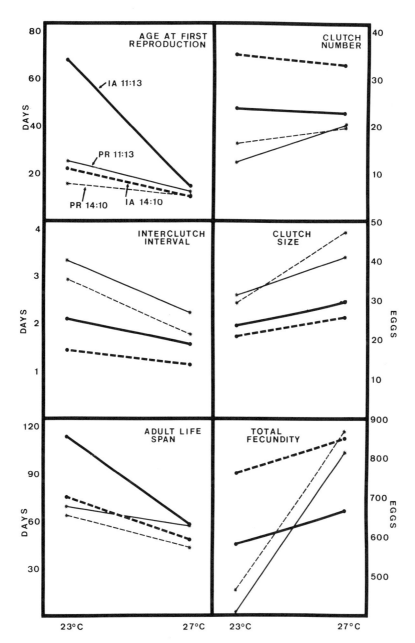

Figure 12-1. Interaction plots showing responses of life history traits in *Oncopeltus fasciatus* populations from Iowa (IA) and Puerto Rico (PR). A difference between points on the same line indicates a response to temperature (e.g., Puerto Rico bugs lay between 40 and 50 eggs in each clutch at 27°C and only 30-35 at 23°C), whereas a difference between lines indicates a response to photoperiod (note, e.g., clutch number for Iowa is greater at LD 14:10). Diverging lines indicate an interaction between temperature and photoperiod as in age at first reproduction (indicating diapause) for Iowa. Note that there are both absolute differences between traits in two geographical source populations (e.g., clutch size) and differences in the way traits respond to environmental variables (e.g., total fecundity).

The Iowa and Colorado populations are subspecifically distinct (Slater and Knop 1969). The Colorado bugs belong to the Western subspecies with an extensive white spot on the wing membrane, whereas the Iowa (Eastern) insects usually lack such spotting. The two populations, however, probably are connected by gene flow because *L. kalmii* is more or less continuously distributed across the Great Plains. The geographical differences in morphology nevertheless suggest the possibility of locally evolved coadapted gene complexes.

With the exception of flight (Caldwell 1974), *L. kalmii* has not been studied with respect to geographical variation in life histories. We therefore had little specific information on which to base predictions as to the outcome of population crosses. We did expect from general considerations (Falconer 1960) and the *O. fasciatus* data that genetic divergence, if it occurred, would result in intermediate expression of hybrids for such traits as size but F_1 overdominance or subviability in fitness characters, such as fecundity. We did observe overdominance in fecundity and related traits so the various lines were carried to the F_2 to see whether breakdown in the hybrids occurred there.

Methods

The two populations of *O. fasciatus* were derived from individuals captured in the field with the stock cultures of each population founded by more than 50 individual genotypes. The stocks were maintained as mass cultures in the laboratory at LD 14:10 and 27°C. The parental pairs used in our crosses were first-generation descendants of the field captured insects. Eight such pairs for each of the eight crosses (see Figure 12-2) were taken from the stock cultures and placed in individual petri dishes with milkweed seed, a water bottle with a cotton wick, and a small wad of cotton for oviposition. The latter was changed daily, and the offspring of each pair were reared separately under the same photoperiod as their parents. Temperature in both short and long days was 24°C by day and 21°C at night for both parental and offspring generations. At adult eclosion the offspring were randomly paired (N = 8 pairs descended from each cross) with each pair placed in a petri dish. Life history traits were then measured for each of the 512 paired females. Further details for rearing *O. fasciatus* in the laboratory under controlled conditions are given in Dingle et al. (1977). A complete description of the methods used here and a full statistical treatment of all results is to appear elsewhere (W. S. Blau, in preparation).

The *L. kalmii* stocks were similarly derived from field captured individuals and were descended from 13 and 18 genotypes from Colorado and Iowa, respectively. Offspring of these founders were the parental pairs used in the initial crosses of 16 matings per line (see Table 12-2). Pairs were reared in petri dishes with milkweed seed, a cotton wick, and dry cotton for oviposition in a LD 16:8 photoperiod; temperatures averaged approximately 30°C at night and 33°C by day. Offspring were reared separately from the parents and at adult eclosion were randomly paired inter se to form the parents of the next generation. Life history traits were determined for each of the paired females in both generations.

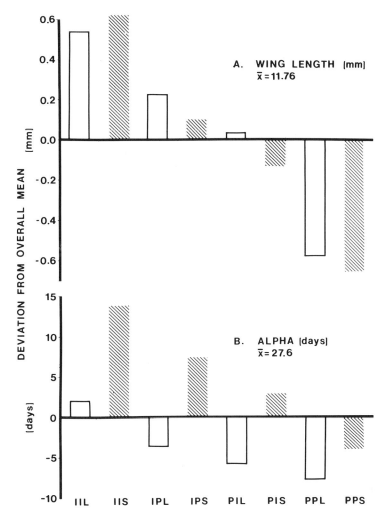

Figure 12-2. Characters (A, wing length; B, age at first reproduction, or alpha) intermediate between parentals in crosses between *O. fasciatus* populations. Mean values for each experimental line are plotted as deviations from the overall mean ($\overline{x}$), e.g., in A larger bugs show upward deviations from the 0.0 line which represents the overall mean of 11.76 mm. Open bars indicate long-day conditions (LD 14:10) and hatched bars, short days (LD 11:13). Crosses and conditions are also indicated below; I, Iowa; P, Puerto Rico; L, long day; and S, short day; females are indicated first in crosses. (For example, IIL indicates "pure" Iowa in long day, whereas IPS indicates Iowa female by Puerto Rico male in short day.)

Results of the Population Crosses: *Oncopeltus fasciatus*

An array of life history traits (Table 12-1, Figures 12-2 through 12-4) was measured on
the eight within- and between-population crosses of *O. fasciatus*. For two of the traits
hybrids resulting from between-population crosses were intermediate in value between
offspring of the original parental populations; this was the case under both long- and
short-day conditions. The first of these is body size. We include size as a life history
character because a number of studies have demonstrated its correlation with classic
life table statistics (Blueweiss et al. 1978, Stearns 1976), and in *O. fasciatus* it is associated with migration (Dingle et al. 1980b). Our measure of body size is wing length,

Table 12-1. Values of z for Mann-Whitney U tests and levels of significance for indicated comparisons within life history traits

Trait	Comparison	z score	p
Wing length (Figure 12-3A)	Iowa vs. Puerto Rico	12.8	<0.001
	Hybrids with Iowa mothers vs. hybrids with Puerto Rico mothers	2.8	<0.005
	Long day vs. short day	2.4	<0.02
Age at first reproduction (α) (Figure 12-3B)	Iowa male vs. Puerto Rico female	7.1	<0.001
	Long day vs. short day	4.5	<0.001
	Hybrids with Iowa mothers vs. hybrids with Puerto Rico mothers	2.2	≅0.02
Interclutch interval (Figure 12-4A)	Hybrids vs. purebreds (long days)	6.5	<0.001
	Hybrids vs. purebreds (short days)	2.5	<0.02
	Hybrids (long days) vs. hybrids (short days)	6.2	<0.001
Reproductive period (Figure 12-4B)	Hybrids vs. purebreds	3.5	<0.001
	Long day vs. short day	1.6	0.2 > p > 0.15
Clutch number (Figure 12-4C)	Hybrids vs. purebreds	6.0	<0.001
Total fecundity (Figure 12-4D)	Hybrids vs. purebreds	3.2	<0.002
Development time (Figure 12-5A)	Hybrids vs. Iowa (short day)	3.4	<0.001
	Hybrids vs. Puerto Rico (short day)	7.2	<0.001
	Hybrids vs. purebreds	5.5	<0.001
Clutch size (Figure 12-5B)	Iowa vs. Puerto Rico	8.4	<0.001
	Hybrids with Iowa mothers vs. hybrids with Puerto Rico mothers	2.3	≅0.02
	Puerto Rico short day vs. Puerto Rico long day	3.2	<0.002

which is highly correlated with weight at adult eclosion (Dingle et al. 1980b). The data for among sample variation in wing lengths is shown in Figure 12-2A. First, it is apparent that Iowa bugs are larger (upward deviation of bars) than Puerto Rico bugs; this difference is statistically significant (z scores and p values for Mann-Whitney U tests of all comparisons are given in Table 12-1). The same difference occurred in earlier samples from the two geographic sources (Dingle et al. 1980b). Second, offspring of interpopulation crosses are intermediate but show a maternal effect; hybrids with Iowa mothers are larger than those with Puerto Rico mothers. Finally, there is an environmental influence on body size with the overall mean wing length of long-day bugs greater than that of short-day insects. Body size, as estimated by wing lengths, therefore is in these bugs a result of gene-environment interactions involving photoperiods.

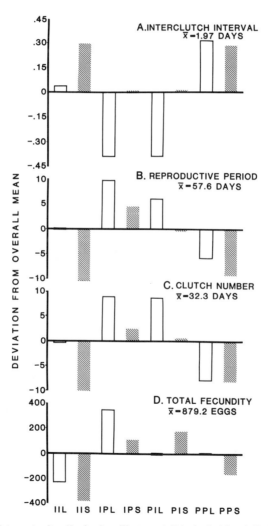

Figure 12-3. Life history traits displaying "heterosis" in hybrids of *O. fasciatus* relative to intrapopulation crosses. Heterosis is indicated in each case by a deviation from the overall mean different from "purebred" deviations. Labeling as in Figure 12-2.

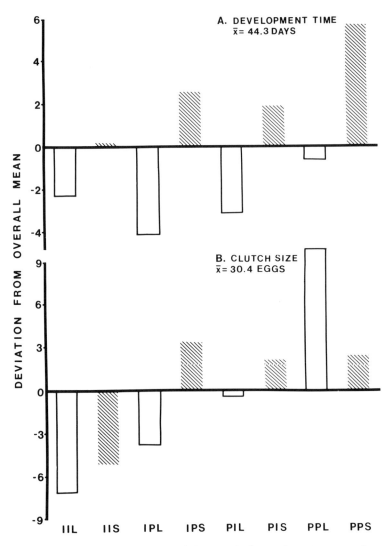

Figure 12-4. Influence of environment (photoperiod) on the outcomes of crosses of *O. fasciatus.* (A) Development time for hybrids is intermediate in short days, but displays heterotic effects in long days. (B) Clutch size is intermediate in long days, but there is apparent dominance deviation toward the Puerto Rico parents in short days. Labeling as in Figure 12-2.

The results for age at first reproduction (α), as measured in days postadult eclosion, are similar to those for wing length (Figure 12-2B). Iowa females take longer to initiate oviposition following adult eclosion than Puerto Rico females, and short days produce delays relative to long days, again demonstrating genic interactions with photoperiod. These data are consistent with earlier results indicating that short days are more likely to induce an adult reproductive diapause in the Iowa than in the Puerto Rico populations (Dingle et al. 1980a). As with wing length, hybrid offspring of interpopulation crosses are intermediate but with maternal influences; bugs with Iowa mothers wait longer before ovipositing than those with Puerto Rico mothers. A word of caution is necessary here, however. Diapause consists both of the initial response to an environmental cue (photoperiod) and physiological changes leading eventually to emergence (Tauber and Tauber 1976). We have evidence that the photoperiodic response in *O. fasciatus* may segregate in relatively simple Mendelian fashion (W. S. Blau, in preparation), and the observed intermediate hybrid α may be a combination of initiation and emergence characteristics. These characteristics will need to be sorted out in future studies of diapause genetics (cf. also Lumme 1978, Vepsäläinen 1978).

Four life history traits in *O. fasciatus*, all relating directly to reproduction, showed "heterosis" in the interpopulation hybrids relative to offspring of intrapopulation crosses within the parental groups. "Heterosis" is used here to mean any hybrid deviation from the overall mean either greater or less than that of the "purebreds." The first of these traits is the average interval between the first 10 clutches of eggs oviposited by each female, i.e., the interclutch interval (Figure 12-3A). Under both environmental regimens, hybrids produced clutches more frequently than purebreds. Among hybrids the interclutch interval was less in long than short days, with the interval in short days being similar to the overall mean, whereas in long days clutches were produced with greater frequency. The influence of day length therefore is substantial.

We measured the duration of the reproductive period as the interval between the first egg clutch and death (which occurred within 2 days of the last clutch); it was longer in the hybrids than in the purebreds (Figure 12-3B). The difference between pooled purebreds was statistically significant (Table 12-1). Unlike interclutch interval, an influence of photoperiod was not demonstrated; bugs reared in the two photoperiodic conditions did not differ significantly in reproductive life span.

The final two traits displaying heterosis were number of clutches and total fecundity (Figures 12-3C, D); these can be considered together. Both are a function of the interclutch interval and the length of the reproductive period. Because interclutch interval is less (Figure 12-3A) and reproductive life span greater (Figure 12-3B) in the hybrids, more clutches and greater fecundity are expected as well. This is indeed the case, with hybrids significantly different from parentals, so that to this extent the traits covary. Heterotic effects for interpopulation crosses are therefore substantial in an array of traits whose net result suggests increased fitness as indicated by higher output of offspring.

In two cases the rearing environment influenced the type of outcome obtained from the *O. fasciatus* population crosses. The first of these is development time measured from the hatching of the egg to adult eclosion (Figure 12-4A). In short days the hybrids are intermediate between the purebreds; the pooled hybrid sample is significantly different from either purebred group. Under long days, however, heterosis was observed in the hybrids whose pooled mean was significantly greater than the pooled purebred mean. The second case of differential environmental influences is that of clutch size (Figure 12-4B). The larger clutches of Puerto Rico females confirm earlier results (Figure 12-1). Field experiments suggested that the clutch size difference might be an adaptive consequence of different predation pressures (Dingle 1981). In the results shown in Figure 12-4B, hybrids are clearly intermediate to the purebred offspring, with hybrid bugs having Iowa mothers producing smaller clutches than those having Puerto Rico mothers. Under short-day conditions clutch size is similar in all

Table 12-2. Means ± S.E. for various life history traits for *L. kalmii* females from Iowa (I), Colorado (C), and interpopulation crosses (female indicated first in cross)

Trait	Generation	I × I	I × C	C × I	C × C
Pronotovertex length (mm)[a]	G_1	3.88 ± 0.07	3.87 ± 0.04	3.85 ± 0.06	3.95 ± 0.04
	G_2	3.73 ± 0.07	3.67 ± 0.04	3.79 ± 0.05	3.64 ± 0.05
Nymphal development time (days)[b]	G_1	17.34 ± 0.25	17.80 ± 0.33	17.21 ± 0.24	17.81 ± 0.19
	G_2	19.25 ± 0.44	19.32 ± 0.42	19.00 ± 0.33	18.86 ± 0.30
Age at first reproduction (days posteclosion)	G_1	4.31 ± 0.16	3.87 ± 0.15	3.64 ± 0.23	3.92 ± 0.16
	G_2	4.05 ± 0.14	4.09 ± 0.29	3.55 ± 0.20	3.82 ± 0.23
Clutch size (eggs)	G_1	34.78 ± 3.17	39.74 ± 3.85	38.74 ± 2.77	45.79 ± 2.62
	G_2	32.31 ± 3.46	32.77 ± 4.78	43.58 ± 4.51	31.85 ± 2.92
Interclutch interval (days)	G_1	0.90 ± 0.03	0.99 ± 0.08	0.96 ± 0.06	1.11 ± 0.07
	G_2	1.29 ± 0.21	1.09 ± 0.05	1.07 ± 0.11	1.00 ± 0.08
Percentage fertility[c]	G_1	87.1	91.5	89.0	94.2
	G_2	86.3	84.1	94.6	95.6
N	G_1	16	15	14	13
	G_2	10	11	10	11

[a]Measured from anterior vertex of head to posterior margin of pronotum.
[b]Interval between laying of egg and eclosion of adult.
[c]Mean percentage of eggs hatching per clutch per female.

groups except the intrapopulation Iowa cross, suggesting a dominance deviation toward the Puerto Rico parents. Analysis of variance indicated significant differences among the short-day samples; a modified test of least significant differences among groups indicated two homogeneous subsets, the intrapopulation Iowa cross and the remainder. Hybrids therefore were more like their Puerto Rico parents with respect to number of eggs per clutch. Puerto Rico short-day bugs, however, produced smaller clutches than their counterparts on long days, suggesting possible gene flow or residual genes from North America (Dingle et al. 1980a). The difference in mean clutch size between Iowa females on long and short days was not significant. Therefore, day length exerts a significant effect on the expression of clutch size and development time, but the outcome for the former with respect to the influence of day length is essentially the reverse of that of the latter (Figure 12-4A), where intermediate offspring occurred in short days.

Results of Population Crosses: *Lygeaus kalmii*

In the case of six life history variables measured in offspring of *L. kalmii* crosses, no phenotypic differences were observed among the four breeding groups (Table 12-2).

For three variables heterosis occurred in the hybrid offspring of interpopulation crosses (Figure 12-5) and was apparent in the second generation as well as the first. This is in contrast to expectations based on similar experiments in *Drosophila* (Wallace and Vetukhiv 1955, but see also Endler 1977), because no breakdown occurred when hybrids were crossed inter se. For each of the three traits reproductive period from the first oviposition to the last, number of clutches, and total fecundity, comparisons of means were made for each generation between grouped hybrid scores and combined parental population scores. The values of z from the Mann-Whitney U statistics with levels of significance are given in Table 12-3.

As was the case in *O. fasciatus*, the traits displaying heterosis in the interpopulation crosses of *L. kalmii* relates to output of offspring. The number of clutches of eggs produced obviously will be a function of the reproductive period unless females with short reproductive lives produce clutches more rapidly. There is no evidence that this is the case here. Again, unless there is a tendency for larger clutches in females with short reproductive lives, fecundity will be a function of reproductive life span. We have found no evidence for such a tendency and conclude that the increased number of clutches and total fecundity of the hybrids from interpopulation crosses is a consequence of the longer reproductive life span of these females. The phenotypic (and presumably genotypic) architecture of this array of traits is therefore apparent.

The Genetic Basis of Differences in Life History Strategies

The evolution of migration and diapause capabilities has enabled the Iowa population of *O. fasciatus* to undertake annual summer invasions of continental North America. These behaviors are essentially absent in the population from Puerto Rico. In the latter case the relatively stable tropical climate of our Vega Alta collection site is favorable for breeding throughout the year, and only local movements are necessary to exploit the fruiting host plants (Dingle 1981). Concomitant with the exigencies imposed by

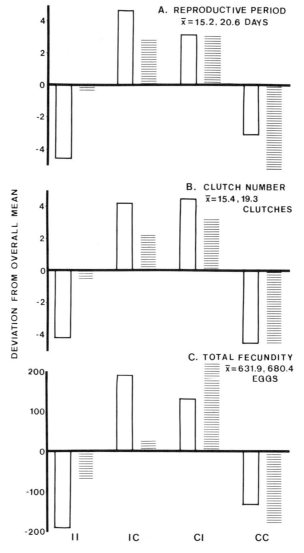

Figure 12-5. Characters displaying heterosis is *Lygaeus kalmii*. Open bars are first-generation offspring and hatched bars are second generation. First number following x̄ = overall mean for first generation and second number is overall mean for second generation. Crosses are indicated below with female parent first (e.g., IC means Iowa female by Colorado male).

climatic and other environmental differences, we should expect the evolution of differences in life histories with appropriate assembling of covariance relationships among traits to construct relevant "strategies." Life history variation between the two geographically separated populations does in fact exist (Figure 12-1), and to the extent that it has been investigated in the field (Dingle 1981), it seems to reflect local adaptive advantage. In North America life cycle patterns presumably have been evolving since the last glaciation, but there is a strong possibility that *O. fasciatus* has been

Table 12-3. Values of z for Mann-Whitney U tests and levels of significance for comparisons between pooled hybrid and pooled parental *L. kalmii* population crosses

Trait[a]	Generation	z	p
Reproductive period (days)	G_1	3.35	<0.001
	G_2	1.48	=0.069
Number of clutches[b]	G_1	3.26	<0.001
	G_2	1.48	=0.069
Total fecundity[c]	G_1	2.94	<0.002
	G_2	1.60	=0.055

[a]Sample sizes as in Table 12-1.
[b]Mean number of clutches produced per female lifetime.
[c]Total number of eggs produced per female lifetime.

present on Puerto Rico only since the European colonization of the Western Hemisphere (Dingle et al. 1980a).

In contrast, the two *L. kalmii* populations occur in similar climatic conditions and are connected by a continuous distribution (and presumably gene flow) across the Great Plains. Differences in local migration patterns do exist between bugs from the two sites (Caldwell 1974), but several life history traits are not statistically distinguishable (Table 12-2). In this chapter we have outlined some of our initial attempts to analyze the extent to which gene differences influence life histories in the two sets of circumstances.

Our analysis using offspring of population crosses indicates that life history traits differ in the manner in which they are expressed in hybrid relative to purebred offspring in geographically separated source populations of the two milkweed bugs. This result demonstrates that life histories may be organized in patterns of covarying traits with an array of genetic structures. These traits evidently differentiate at dissimilar rates so that hybrids may display assorted combinations of phenotypes depending on the nature of their underlying gene influences and on the environment in which they are expressed (e.g., Figure 12-4). These between-population interactions are potentially important in those kinds of natural situations where hybrids may occur in environments unlike those preferred by either parent. Such a case might lead to "parapatric" speciation (Endler 1977). Where populations are sufficiently isolated, life history and other differences could evolve to produce allopatric speciation (Mayr 1963). Possible modes of population differentiation revealed by crossing populations are outlined in Figure 12-6 and can serve as a focus for further discussion of genes and life histories.

For some traits neither source populations nor the hybrids derived from them show detectable differentiation. This result was observed in *L. kalmii* (Table 12-2) and has also been reported for other species. For example, no significant differences were found for body size or development time among hybrids and purebreds of several populations of *Drosophila subobscura* taken in a transect across southern Scotland (McFarquhar and Robertson 1963). These results do not necessarily mean that no gene differences for the traits occur among the sampled populations of either *L. kalmii* or *D. subobscura*, but only that they were not detected by the particular crosses. Modifier genes in appropriate combinations can lead to this outcome even if other gene differences exist

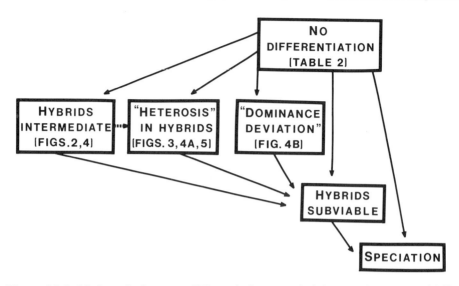

Figure 12-6. Modes of character differentiation revealed by crossing geographially separated populations of the same species. Speciation can occur via several pathways. Those modes observed in this study are indicated by references to appropriate tables and figures, whereas the remaining two are evident from the literature (e.g., Tauber and Tauber 1977, Oliver 1979, Templeton 1979). Dotted arrow indicates that additional transitions are also possible. Various modes of character differentiation can occur in the same populations at the same time.

(Endler 1977). Similarities within and among regional samples would be expected before differing selection pressures or drift resulting from spatial or temporal segregation produced population differentiation. The absence of such homogeneous traits in the *O. fasciatus* crosses indicates significant differentiation between the source areas examined.

Intermediate expression of traits in hybrids is represented in *O. fasciatus* by wing length and age at first reproduction (Figure 12-2) and by development time and clutch size in certain environments (Figure 12-4). This mode has also been observed in *D. subobscura* for both body size and development time in regional crosses between Scotland and Egypt (McFarquhar and Robertson 1963) and in many insect species for diapause-related traits (Danilevskii 1965, Dingle et al. 1977, Hoy 1978). Because age at first reproduction in *O. fasciatus* reflects diapause in response to photoperiod, the result is not surprising. We have independent evidence for Iowa bugs that there are high proportions of additive genetic variance for photoperiodic response (Dingle et al. 1977) and wing length (Hegmann and Dingle, Chapter 10, this volume). These data and the results from our crosses (and others) suggest that high additive genetic variance in the parent populations leads to intermediate expression of the traits in the offspring, just as if parents were drawn from the same source. The selective regimes acting on these traits apparently allow the underlying gene complexes to remain "open" (Carson 1975) rather than to become specifically coadapted.

No current quantitative genetic theory seems adequate to deal with such between-population additive genetic variance although simpler models based on different types of modifiers can explain it (Endler 1977). This type of variance seems to occur when there is no consistent selection for a particular adaptive response or when a species must scan an array of shifting environments, e.g., over a geographical range (Dingle et al. 1977, Hoffman 1978, Istock 1978). This could lead to an "open" genetic system in Carson's (1975) sense, and analysis of the relation between openness and type of genetic variance (and modifiers) suggests a direction further theoretical development might take. Needed are studies with independent estimates of additive genetic variances in each of two (or more) parental populations which are then crossed and genetic variances estimated from the crosses. Insect photoperiodic response should be an excellent candidate trait because of the frequent occurrence of hybrid intermediacy as cited above. The oft-noted presence of maternal influences indicates a further additive environmental component. Life history variables influenced by high levels of additive genetic variance may be important in the early stages of population differentiation especially if they result in life cycle allochronism. It is perhaps also worth noting that in one well-studied system, *Drosophila mercatorum*, genetic diversity predicted the a priori chances for a successful "genetic revolution" (Templeton 1979 and Chapter 5, this volume).

   Differentiation in which offspring traits from interpopulation crosses expressed "heterosis" relative to those from parental crosses was displayed in both *O. fasciatus* and *L. kalmii* by a character complex which combines to yield total fecundity (Figures 12-3 and 12-5). This complex includes reproductive period, number of clutches, and in *O. fasciatus* interclutch interval. In the latter species development time, which can influence the timing of reproduction, displays environment specific heterosis (long days). Because fecundity is a direct measure of offspring production, the complex bears a close relation to fitness. For this reason, it is interesting that heterosis occurs in the expression of these traits, for this result presumably reflects differential coadaptation (Endler 1977) specific for the selective regimes experienced by the parental populations. Such differential coadaptation depends on the behavior of both major genes and the types of modifiers (Endler 1977), and the genetic correlation among the traits is an important subject for quantitative genetic analysis as stressed in this symposium (e.g., Hegmann and Dingle, Chapter 10, and Lande, Chapter 2, this volume). An interesting question that arises here is can we predict which traits are most likely to show differential coadaptation in populations of various geographical origins and thence discern something of their genetic structure? In *L. kalmii* these fecundity-related traits also showed heterosis in the F_2 offspring of inter se crosses within hybrid lines so that the breaking apart of presumptive coadapted gene complexes was not demonstrated, again possibly as the result of appropriate modifier combinations (Endler 1977).

   The expressions of two traits in the *O. fasciatus* hybrid offspring were environment dependent; these were development time and clutch size (Figure 12-4). The former was intermediate between the purebred offspring in short days (presumably reflecting additive genetic variance) but displayed heterosis in long days (presumptive coadaptation). The latter displayed directional dominance (Figure 12-6) in short days, whereas it was intermediate in the hybrids in long. Development time was not different between purebreds and hybrids in *L. kalmii* (Table 12-2), was intermediate in *Drosophila subobscura* in the environment tested (McFarquhar and Robertson 1963), and showed sex differences in Lepidoptera (Oliver 1979). It displayed relatively high additive genetic variance in a long-day Iowa population (Hegmann and Dingle, Chapter 10, this volume). If these comparative results do not simply reflect bias from the small number of species sampled, the variety evident in development time makes this trait an interesting one to examine with respect to the genetics of population differentiation.

The following additional points are worth noting with respect to the various modes of expression. First, no significant differences in parental phenotypes were detected in some traits displaying heterosis in the hybrids. Therefore, the heterotic effects indicate that the same parental phenotypes are the products of different genotypes, revealed only when the hybrids express phenotypes different from either parent. Second, the kind of expression, whether heterosis, dominance deviation, or hybrid intermediacy, may be environment specific, indicating that coadaptation, to the extent that it occurs, involves gene-environment interactions. The covariance of life history trait and environment in the coadaptation process therefore is indicated. Finally, the covariation in a life history character complex, such as that combining in fecundity, emphasizes the significance of the relation between genetic coadaptation and covarying phenotypic traits.

Differentiation between populations can lead to speciation by a number of possible routes (Figure 12-6). It can proceed directly by means as varied as life cycle allochronism from dominance effects at a few loci (Tauber and Tauber 1977 and Chapter 4, this volume) or major coadaptations apparently involving major portions of the genome (Templeton 1979 and Chapter 5, this volume). However, there can be intermediate steps proceeding via one or more characters passing through one or more modes as Figure 12-6 indicates. Whatever the speciation process, the importance of complex life history adaptations is fundamental because these are constructed of covarying and genetically correlated (Hegmann and Dingle, Chapter 10, this volume) arrays of characters profoundly influenced by developmental rates and events. The phenotypic structure of the array can determine the compatability of populations of differing origin. The role of complex life history adaptations therefore is apparent, and understanding their genetic structure is an important goal of evolutionary biology.

An important conclusion from the preceding discussion is that the various modes of character differentiation can occur in the same species at the same time. A possibility that emerges as a consequence is that the coadaptations underlying different traits may play key roles at different points along the continuum of population divergence. If so, a major question is whether the timing or staging of these roles is consistent across particular taxa, or whether there is predictable diversity. This and related questions of effective differences between populations of various species are very much in need of analysis (McFarquhar and Robertson 1963, Endler 1977). The dynamic view of population structure presented here implies long-term cumulative efforts by evolutionary biologists to understand the genetic and environmental bases of divergence. Life histories should be an important focus for these efforts.

Summary and Conclusions

Populations of the milkweed bugs *Oncopeltus fasciatus* and *Lygaeus kalmii* contain gene-based differences in life history traits as determined by comparisons among offspring of intra- and interpopulation crosses. Traits that do not differ phenotypically may still be the result of different genotypes.

In differentiating populations, traits may vary with respect to modes of differentiation. The rate of divergence is limited by the heritability of the trait and is directed by the genetic variance-covariance structure. Both the genetic structure and the process of differentiation are dynamic.

The kind of expression can be environment specific. Coadaptation in the genome thus involves interactions with differing environments, further emphasizing the dynamic genetic structure.

Coadaptations underlying different traits may vary in importance in the process of population differentiation. A major question for further research is: Can it be predicted which traits will be important and in what circumstances from their variance-covariance structure? Answering that question should yield significant insights into life history evolution and speciation.

Acknowledgments. Patricia Blau, Lori Cushman, Edward Klausner, and Elizabeth R. Miller have contributed excellent technical assistance to this study. The manuscript was thoroughly reviewed by J. David Allan, John A. Endler, and Charles G. Oliver, to all of whom we are most grateful. Supported by grants from the U.S. National Science Foundation to H. D. and a Postdoctoral Traineeship to W. S. B. from the National Institute of Mental Health.

References

Bachmann, K.: Temperature adaptations of amphibian embryos. Am. Nat. 103, 115-130 (1969).

Bell, G.: The costs of reproduction and their consequences. Am. Nat. 116, 45-76 (1980).

Blueweiss, L., Fox, H., Kudzma, V., Nakashima, D., Peters, R., Sams, S.: Relationships between body size and some life history parameters. Oecologia 37, 257-272 (1978).

Caldwell, R. L.: A comparison of the migratory strategies of two milkweed bugs, *Oncopeltus fasciatus* and *Lygaeus kalmii*. In: Experimental Analysis of Insect Behaviour. Barton Browne, L. (ed.). New York: Springer-Verlag, 1974, pp. 304-316.

Carson, H. L.: The genetics of speciation at the diploid level. Am. Nat. 109, 83-92 (1975).

Danilevskii, A. S.: Photoperiodism and Seasonal Development of Insects. Edinburgh: Oliver & Boyd, 1965.

Dingle, H.: Life history and population consequences of density, photoperiod, and temperature in a migrant insect, the milkweed bug *Oncopeltus*. Am. Nat. 102, 149-163 (1968).

Dingle, H.: The experimental analysis of migration and life history strategies in insects. In: Experimental Analysis of Insect Behaviour. Barton Browne, L. (ed.). New York: Springer-Verlag, 1974, pp. 327-342.

Dingle, H.: Migration and diapause in tropical, temperate, and island milkweed bugs. In: Evolution of Insect Migration and Diapause. Dingle, H. (ed.). New York: Springer-Verlag, 1978, pp. 254-276.

Dingle, H.: Geographic variation and behavioral flexibility in milkweed bug life histories. In: Insects and Life History Patterns: Geographic and Habitat Variation. Denno, R. F., Dingle, H. (eds.). New York: Springer-Verlag, 1981, pp. 57-73.

Dingle, H., Alden, B. A., Blakely, N. R., Kopec, D., Miller, E. R.: Variation in photoperiodic response within and among species of milkweed bugs (*Oncopeltus*), Evolution 34, 356-370 (1980 a).

Dingle, H., Blakley, N. R., Miller, E. R.: Variation in body size and flight performance in milkweed bugs (*Oncopeltus*). Evolution 34, 371-385 (1980 b).

Dingle, H., Brown, C. K., Hegmann, J. P.: The nature of genetic variance influencing photoperiodic diapause in a migrant insect, *Oncopeltus fasciatus*. Am. Nat. 111, 1047-1059 (1977).

Dobzhansky, T.: Genetics of the Evolutionary Process. New York: Columbia University Press, 1970.

Endler, J. A.: Geographic Variation, Speciation, and Clines. Princeton: Princeton University Press, 1977.

Falconer, D. S.: Introduction to Quantitative Genetics. Edinburgh: Oliver and Boyd, 1960.

Frazetta, T. H.: Complex Adaptations in Evolving Populations. Sunderland, Mass.: Sinauer, 1975.

Hoffman, R. J.: Environmental uncertainty and evolution of physiological adaptation in *Colias* butterflies. Am. Nat. 112, 999-1015 (1978).

Hoy, M. A.: Variability in diapause attributes of insects and mites: Some evolutionary and practical implications. In: Evolution of Insect Migration and Diapause. Dingle, H. (ed.). New York: Springer-Verlag, 1978, pp. 101-126.

Istock, C. A.: Fitness variation in a natural population. In: Evolution of Insect Migration and Diapause. Dingle, H. (ed.). New York: Springer-Verlag, 1978, pp. 190.

Kruckeberg, A. R.: Variation in fertility of hybrids between isolated populations of the serpentine *Streptanthus glandulosus* Hook. Evolution 11, 185-211 (1957).

Lande, R.: Quantitative genetic analysis of multivariate evolution, applied to brain: body size allometry. Evolution 33, 402-416 (1979).

Lewontin, R. C.: The Genetic Basis of Evolutionary Change. New York: Columbia University Press, 1974.

Lindsay, D. W., Vickery, R. K., Jr.: Comparative evolution in *Mimulus guttatus* of the Bonneville Basin. Evolution 21, 439-456 (1967).

Lumme, J.: Phenology and photoperiodic diapause in northern populations of *Drosophila*. In: Evolution of Insect Migration and Diapause. Dingle, H. (ed.). New York: Springer-Verlag, 1978.

Mayr, E.: Animal Species and Evolution. Cambridge, Mass.: Harvard University Press, 1963, 797 pp.

McFarquhar, A. M., Robertson, F. W.: The lack of evidence for coadaptation in crosses between geographical races of *Drosophila subobscura* Coll. Genet. Res. 4, 104-131 (1963).

Nichols, J. D., Conley, W., Batt, B., Tipton, A. R.: Temporally dynamic reproductive strategies and the concept of r- and k-selection. Am. Nat. 110, 995-1005 (1976).

Ohno, S.: The preferential activation of maternally derived alleles in development of interspecific hybrids. In: Heterospecific Genome Interaction. Defendi, V. (ed.). Philadelphia: Wistar Institute Press, 1969.

Oliver, C. G.: Genetic and phenotypic differentiation and geographic distance in four species of Lepidoptera. Evolution 26, 221-241 (1972).

Oliver, C. G.: Genetic differentiation and hybrid viability within and between some Lepidoptera species. Am. Nat. 114, 681-694 (1979).

Slater, J. A., Knop, N. F.: Geographic variation in the North American milkweed bugs of the *Lygaeus kalmii* complex. Ann. Entomol. Soc. Am. 62, 1221-1232 (1969).

Sokal, R. R., Taylor, C. E.: Selection at two levels in hybrid populations of *Musca domestica*. Evolution 30, 509-522 (1976).

Solbreck, C.: Migration, diapause, and direct development as alternative life histories in a seed bug, *Neacoryphus bicrucis*. In: Evolution of Insect Migration and Diapause. Dingle, H. (ed.). New York: Springer-Verlag, 1978, pp. 195-217.

Stearns, S. C.: Life history tactics: A review of the ideas. Q. Rev. Biol. 51, 2-47 (1976).

Stearns, S. C.: The evolution of life history traits. Ann. Rev. Ecol. Syst. 8, 145-172 (1977).

Tauber, C. A., Tauber, M. J.: A genetic model for sympatric speciation through habitat diversification and seasonal isolation. Nature (London) 268, 702-705 (1977).

Tauber, M. J., Tauber, C. A.: Insect seasonality: Diapause maintenance, termination, and postdiapause development. Ann. Rev. Entomol. 21, 81-107 (1976).

Taylor, L. R., Taylor, R. A. J.: Aggregation, migration, and population mechanics. Nature (London) 265, 415-421 (1977).

Taylor, L. R., Taylor, R. A. J.: The dynamics of spatial behavior. In: Population Control by Social Behaviour. Ebling, F. J., Stoddart, D. M. (eds.). London: Institute of Biology, 1978.

Templeton, A. R.: The unit of selection in *Drosophila mercatorum*. II. Genetic revolutions and the origin of coadapted genomes. Genetics 92, 1265-1282 (1979).

Vepsäläinen, K.: Wing dimorphism and diapause in *Gerris*: Determination and adaptive significance. In: Evolution of Insect Migration and Diapause. Dingle, H. (ed.). New York: Springer-Verlag, 1978, pp. 218-253.

Vetukhiv, M.: Fecundity of hybrids between geographic populations of *Drosophila pseudoobscura*. Evolution 10, 139-146 (1956).

Vetukhiv, M.: Longevity of hybrids between geographic populations of *Drosophila pseudoobscura*. Evolution 11, 348-360 (1957).

Wallace, B.: Inter-population hybrids in *Drosophila melanogaster*. Evolution 9, 302-316 (1955).

Wallace, B., Vetukhiv, M.: Adaptive organization of the gene pools of *Drosophila* populations. Cold Spring Harbor Symp. Quant. Biol. 20, 303-309 (1955).

PART SIX

Closing Discussion

Following presentation of the formal papers for the conference, those in attendance gathered for discussion of a set of issues that had been selected by Stevan Arnold, Barbara Schaal, and Michael Lynch. We are indebted to these three for their help and to all of those in attendance for contributions to the lively discussion, which we present below as transcribed from tape. We accept responsibility for errors in transcription and we hope the discussion is sufficiently preserved to provide as effective a conclusion for this volume as it did for the conference.

ARNOLD: We want to focus on three major topics in this discussion: (1) The practice and theory of quantitative genetics (Arnold), (2) phenotypic fitness in relation to environment (Lynch), and (3) trade-offs (Schaal).

Study of the evolution of life history traits must be concerned with their genetics. Usually, we assume they are polygenic traits and therefore appropriate subjects for quantitative genetics. Quantitative geneticists, however, would react with horror to traits the life historian would present for analysis. These are traits most intractable to quantitative genetics.

To begin, I shall direct two questions to Dr. Rose and Dr. Lande. (1) How seriously is the methodology of quantitative genetics affected by confounding factors, e.g., maternal effects, genotype-environment interactions, genotype-environment correlations, partial inbreeding, and fluctuations in the environment in which measurements are taken? (2) What are current and future directions for theory and empiricism?

MICHAEL ROSE (University of Wisconsin):

Drosophila melanogaster is a good organisms for this question because it is extremely badly behaved. If you vary or fail to control almost anything, experiments will not work. Some artifacts are maternal effects and variable numbers of progeny per female. It is important to control maternal effects and genetic correlations because the numbers of eggs per female is partly a property of the genes she carries. If you do not control egg production for parent-offspring correlation in sibling analysis, egg production results could be almost meaningless. If the analysis is in different environments from the one in which the population reached evolutionary equilibrium, results would also have almost no meaning. So, it gets complicated.

There is a note of optimism on all of the above, however. Although traits are bad in some organisms, one can control for these at least one at a time in separate experiments. So artifacts do not rule out quantitative genetics. Quantitative geneticists say you cannot use characters that are not strictly neutral—only because you need to presume strict linkage equilibrium with no epistasis and with Hardy-Weinberg equilibrium. If the population is at an evolutionary equilibrium with no linkage or extensive epistasis then many of the results still hold good.

ARNOLD: Do you have comments on the direction quantitative genetics should take?

ROSE: The theory is terrible quantitatively but very useful qualitatively. Lots of interesting problems for experiments are generated by the papers of Lande and Charlesworth.

It seems best to either formulate testable theory or test current theory rather than formulate broadscale theory, erecting models which are difficult or impossible to test.

ARNOLD: We must deal with evolutionary trajectories by a matrix of additive genetic variance and covariance for a suite of traits. The problem is the matrix probably is not constant through time. What is the optimism for theory when as ecologists we worry that critical genetic parameters are wobbling all over the place in nature?

LANDE: I should first point out the difference between doing quantitative genetics in controlled laboratories versus in natural populations. Many complicating assumptions of the field can be overcome with properly designed lab experiments, e.g., problems of genotype-environment correlations which are usually set to zero by randomizing the lab environments.

Genotype-environment interactions can be assessed in controlled conditions, e.g., Dr. Bradley found no interactions with regard to temperature stress. In nature, populations can be tested for the existence of these types of effects by doing certain kinds of experiments, such as cross fostering—at least that will test for genotype-environment correlations to some extent. These sometimes can be difficult to distinguish from maternal effects, but as in artificial breeding experiments, if one simply looks at a number of different types of relatives, including cross correlations between relatives of opposite sex, it should be possible to separate out many of these complicated sources of variance.

As far as environmental changes acting through genotype-environment interactions affecting the structure of the genetic variance-covariance matrix, I think there is very little that can be said as a theoretical point at present on this. It simply must be decided for each individual case how important it may be, and this generally must involve laboratory experiments, I think.

With respect to a strategy for doing theory as compared with a strategy for investigating questions empirically, my approach in developing these simple theories is that at first theories of this kind are useful generally just to give one an intuitive idea how to order the most important things to investigate empirically. One must start with simple cases and treat complications as sorts of deviations from norms. So I do not feel at all guilty about assuming stable age distributions and weak selection when discussing life history evolution at this point because any more complicated or realistic situations will have to be understood in terms of the simpler cases.

ARNOLD: At the end of your talk . . . you outlined a prospect for reconstructing the pattern of selection acting on life history traits. . . . I wonder if you could go over that again for us, since that seems an obvious direction for research.

LANDE: The basic idea would be to compare two contemporary populations which are closely related—like subspecies or congeneric species having homologous life histories—but which differ in some significant way. Then a hypothesis would be constructed from say, purely ecological data as to what selective forces had led to this difference in life histories. This coul apply to any self-contained or independently evolving set of characters, not just life histories. The basic theory assumes a multivariate normal distribution of the characters in both populations and that at least on some transformed scale of measurement one can also assume that additive genetic variances and covariances within the populations have stayed the same through time. Then, if this is true, the formulas I constructed for response to selection in the vector of characters which may represent the life history of the organism are given by this life history variance-covariance matrix times a gradient of selective forces. Even though the selective forces may be changing simply as a result of evolution in the population or as a result of environmental changes, if it is empirically found that the pattern of genetic variances and covariances remains roughly constant (the degree of roughness will have to be decided empirically also), then this equation can be simply extrapolated through time to yield an expression that will give the net forces of selection acting directly on the characters in terms of the partial regressions of the characters on fitness that have acted through time to create the observed difference in the two populations. This is a completely independent methodology for testing at least the magnitude of selective forces that have been acting on life history characters. Now this only gives you relative magnitudes of selection, especially when one does not have an absolute time scale for the divergence of the populations. So this does not really help to construct the actual ecological processes leading to those selective forces, but at least it allows one to test whether the magnitudes one was referring to in the specific ecological model are consistent with this independent genetic analysis. If one then begins to subdivide total fitness into its various components, it might be possible to assign particular products to certain estimates selectively.

HEGMANN: Is it possible to do it the other way around—to argue backward that given the particular genetic variance-covariance structure that exists in a population today, then invoking the assumption that it will remain stable in the face of the selective behavior, is it possible to walk into, say, field A and field B and hypothesize that humidity, temperature, food density, etc., constitute the vector of environmental agents that are a hypothetical source of selection? Can one in any way test whether that hypothesis is right or wrong?

LANDE: At present the theory is only framed rather abstractly with respect to the environment, as are most population genetic models. The only thing explicit in the theory is fitness or l_x and m_x schedules of different phenotypes so it is not possible to work backwards.

HEGMANN: I am a little sensitive to this because I have argued that there are a number of environmental events that could have occurred in the past history of the species to generate any residual variance-covariance matrix that you see any any time. Therefore speculations regarding past historical events that may have left the population with that variance-covariance structure must remain speculations; the power of knowing the genetic variance-covariance structure lies in the promise that if selection is brought to bear in a given way, the population will respond in a given way. That is an artificial selectionist's way of adding confusion to natural selection. Will it ever be possible to incorporate aspects of the environment in some general theory?

LANDE: I think that to do that would require an exploration in a number of fields that have barely been opened at this point. In particular, there is the idea of a norm of reaction of genotypes which goes back to the question of genotype-environment interactions. Dobzhansky has discussed this quite a bit and there have been a few experiments looking at range of phenotypes expressed by certain genotypes in a range of environments and doing this at a population level. That is the information that would be required to look for genetic changes occurring in a population in a changing environment. Until one is able to specify ranges of environments and norms of reactions in that way, I do not think it will be possible to make an accurate estimation.

MICHAEL WADE (University of Chicago)

I think it would be extremely useful to empiricists if we could get a sensitivity analysis of some of these models. If you argue on theoretical grounds the estimation of the genetic variance-covariance matrix and knowledge of population mean vector differences between the populations will allow you to reconstruct some hypothesis about the selective history; that is a material thing. Then we know, from many of the talks presented here, that measures on a thousand individuals may yield standard errors that are quite large. How do we know when it is worthwhile making measures on another thousand individuals? How effectively will these additional observations increase the accuracy in reconstructing selective past history? So a sensitivity analysis on these models would be a great help to the empiricist in designing experiments.

ANTONOVICS:

These are all good points but we should avoid getting hung up in the "politics" of quantitative genetics. The basic point, the thing you do not want to do, is use phenotypic correlations as the basis of predictions. That has been the tempting thing for ecologists to do and it is the wrong thing. In a quantitative sense you can show that the phenotypic correlation is often in a different direction (sign) from the genetic correlation. Just showing that is often a very powerful thing; and its very easy to do. In an experimental situation you can keep individuals as separate sibships—as half-sibships or full sibships. A sibship variance-covariance matrix can be estimated regardless of whether its maternal or genetic and that alone will tell you about the genetic relationships. You can dissect out various sibship components of variance—I call it "genetics without tears." This can avoid some of the concerns with sampling—instead of sampling at random you sample in a family structure.

Mike Wade's point is well taken. As a graduate student Mike Ross did a simulation study. The presence of a suite of characters with a genetic correlation as low as -0.20 could slow the rate of evolution (of an index trait) tremendously. In order to measure a genetic correlation of -0.20 as significant requires, I think, a sample size of 200-300.

ARNOLD: Still, it is not as if we are not interested in phenotypic correlations. As Dr. Doyle has pointed out, they still play an important role in structuring the selection that acts on characters. The important point is that genetic correlations add a whole new dimension to the problem.

ANTONOVICS:

Yes, I think that to deal with correlations in terms of selection they have to translate into genetic terms.

ROSE: While we are on the subject, let me point out that there are good
               reasons for expecting negative genetic correlations among pheno-
               typically positively related fitness components. The point that Dr.
               Antonovics has made is of general significance.

HEGMANN: In that same line and in response to something Dr. Rose said earlier,
               it is true that one could adopt the attitude that maternal effects or
               genotype-environment interactions (or correlations) are nuisance sources
               of variance contrived by some force to confound and confuse quanti-
               tative geneticists. I think that would be a mistake. Things like maternal
               effects in placental mammals provide the opportunity for feedback,
               which is physiologically and functionally very important. The use of
               genetics allows identification of prenatal or postnatal maternal effects
               and that finding alone may be very important. I think it would be a
               mistake to get depressed either with "nuisance" sources of variance
               or with the difficulties of achieving low standard errors. If low but
               important genetic correlations are estimated, they can be tested in short-
               term artificial selection experiments.

LANDE: I think that is an important point in regard to Dr. Wade's comments on
               sensitivity analysis. Standard errors for genetic variances and covariances
               estimated from correlations between relatives are usually much larger
               than those for estimates from realized responses to selection accumu-
               lated over several generations. To predict with a certain degree of accu-
               racy the multivariate response to selection, one should get a preliminary
               estimate of the genetic variance-covariance matrix and determine which
               were the most sensitive elements for the reconstruction. Increased sam-
               pling efforts should be aimed at those particular characters and a selec-
               tion experiment could be aimed at that part of the matrix.

DOYLE: I am bothered with the sensitivity problem. If one undertakes these
               laborious genetic experiments with the hope of answering some of the
               important questions, such as why is there so much genetic variance
               when the simplest theory suggests it should disappear? Or, why does
               the fossil record make it appear that animals spend so much time not
               evolving, rather than evolving? Or, is disruptive selection on fitness
               traits an important part of speciation? These questions are being ad-
               dressed by empirical studies on the assumption that they can be answered
               at the genetic level. The alternative, which I discussed earlier, is that the
               phenotypic relations among traits may be such that selection is usually
               essentially zero. What is going on at the genetic level may be essentially
               moot.

LYNCH: The next general question is aimed at the evolutionary ecologists in
               the audience. Can we forget about genetics and talk about the evolution

of life histories to predict the direction life history should evolve in? That is what we have been doing for years. In addition, if we set biological constraints or rules, are there ways of relating the environment to the evolution of life histories specifically? Are there ways of relating phenotypes and their fitness contributions to the environment? This is evolutionary ecology though their theories are couched in general terms.

LANDE: I would like to point out that without considering genetics one can only make a static theory of life history evolution. You could model an optimal life history for a certain environment but in terms of predicting selective forces which had led to differences between two populations, the genetics is controlling the response to selection; so I think that without genetics you can only predict a static optimum.

LYNCH: Can we predict the direction in which life histories will evolve? Are events like "bottlenecks" so important that we cannot do that?

LANDE: The direction of the response to selection on a multivariate complex of life history traits depend critically on the pattern of genetic covariance among them. If a set of traits is completely determined by a given set of genes (i.e., completely pleiotropic or completely genetically correlated) then it is impossible for them to evolve in separate directions, regardless of the forces of selection acting on them. If the correlation (genetically) is not complete but is very high, then it will take a very long time for any evolutionary optimum to be achieved. It is a question how successful any optimization theory can be if it ignores these genetic complications and genetic structures. From the standpoint of dynamics it is the question of how close populations usually are to the optima toward which they are selected. The extent of success in evolutionary ecology may be a measure of that.

ISTOCK: I agree with Dr. Lande, but there may be an antecedent problem. Before anyone can enter the logic of Lande's work, they first have to decide what's to be included in the Z vector (of life history trait means). No member of this group could tell you what entries to include for any species whatsoever. We are going to have to study the organisms as the ecologists do, to decide which features have any contact with important potential selective forces. Unfortunately there is another problem which has slipped by us. We need to find some way to connect that choice of characters to the selective gradient. We have to have a machinery. The answer to the question is that no ecologist has ever gotten a significant result without genetics; but the reverse is also true. It is not at all clear, with the exception of a few qualitative characters, that population geneticists have ever gotten a significant result reasoning from P's and Q's, either!

DINGLE: We also need to be careful to consider the behavior of organisms. The tendency to talk about the environment handling the organisms in a willy-nilly way should be tempered by considering what the organism can do to alter its environment. It is not just a question of whether the ecologists can judge the environmental pressures but how effectively the organism can choose its environments. The important point is that organisms can and do make choices; behavior allows them to make those choices. Therefore the relation between genetics and behavior is important.

MEAGHER: I do not think it is completely fair to say that evolutionary ecologists have been ignorant of genetics. An evolutionary ecologist deals with an organism in its environment as an integrated system which is subject to specific internal constraints peculiar to that organism. Those constraints are going to be governed by genetics.

LYNCH: That leads into the question of trade-offs.

WADE: We have heard several speakers present empirical data concerning changes from one season to the next or one locality to the next in what they consider to be biological constraints in ecological models. One good example was in Ai-chu Wu's poster showing that in one species of *Tribolium* trade-offs exist among a series of life history characters while in another species of the genus they do not in exactly the same set of environmental conditions. The notion that you can perceive some universal set of constraints and thereby build universal models does not seem to hold up. This was demonstrated in the *Oncopeltus* work and in Dr. Giesel's work as well as in *Tribolium*. I do not know of any set of life history characters for which everyone has found the same correlations.

TEMPLETON:

Genetics, of course, does apply to evolutionary processes to a very great extent. It is not simply a side constraint that you can put on an ecological model. In a very real way genetics is where the action is.

Another point from the *Drosophila* work is that we do not always have to treat the genetics as a black box. With the studies of parthenogenesis we found that natural populations were polymorphic for a number of very important life history traits. They were polygenic, but the structure of the polygenic relationships was not as complicated as we had originally assumed. We almost always had one or two major loci with several small modifiers. Other *Drosophila* work on *D. heteroeura* and *D. sylvestris* which differ grossly in morphology demonstrated a major X-chromosome effect operating as a major regulatory gene on several autosomal modifiers. Insecticide resistance has also been studied. Of 217 or so cases, something like 213 have a major locus, once again with modifiers. So it seems to be possible in at least some cases to discover common genetic patterns underlying certain traits. Once you have that you can start making generalizations even if you do not know the details of a given species. There is common genetic architecture and matrices no longer need to be black-box matrices. We can actually start making some assumptions about what the genetic structure should be. That should go a long way, I should think, toward uniting population genetics with ecological genetics.

DOYLE: I would like to suggest that there may be a counter to what Dr. Templeton had said. If you choose to deal with systems that can be easily explained genetically, particularly by major genes, you may be introducing the same kind of bias that was inherent in population genetics when all it could deal with was visible polymorphisms. In other words, it is not a generalization.

TEMPLETON:
As I said, we shall have to approach this empirically. In both the insecicide resistance cases and the *D. mercatorum* life history examples, we have a great deal of variation almost all of which has a similar genetic architecture. And, in much of the hybridization work with rice almost all the traits looked at have had one or two major genes involved with no more than 10 modifiers.

DOYLE: That is true. Many agricultural cases of domestication, for example, have had major genes involved early on in the process.

ANTONOVICS:
I would like to second that. I think evolutionary ecologists make generalizations based on broad environments and broad taxonomic groups. We do not have enough comparative information on correlation structures within one genus or on how correlation structures vary when you put an organism into a class of environments. Another case in point is work showing that during the evolution of crop plants the kind of yield-seed number-seed size relationships follow a fixed trajectory essentially determined by initial constraints. So, you can often break down yield into its various components, but they follow rather straight trajectories. I would argue that we need to do two kinds of things: (1) study these things between different taxa and (2) analyze closely related species within certain of these taxa. We should do experiments predicated on what Dr. Giesel was doing, taking small organisms from one environment to another specific environment (ideally these environments should have ecological meaning) and ask how the variance-covariance structure changes and try to get a feel for how these things happen.

HEGMANN: I think one of the exciting things Dr. Templeton said was that we may be able, in some systems, to fix one or two key loci and contrast the genetic variance-covariance structure in stocks segregating at those loci to that in fixed stocks. We may find key loci that are "playing" with the nature of genetic correlations. This would turn the whole thing around into an analytical scheme where you can say, yes, there are genes that can affect life history structures in a multivariate sense and say that with fairly small population sizes.

MEAGHER: If this situation is the norm—one major gene with several minor ones—does the multivariate normal still hold? That is not the assumption on which quantitative genetics is based, is it? Is this a problem?

LANDE: Yes. Standard theory in quantitative genetics developed by R. A. Fisher actually assumed that there were an infinite number of loci affecting characters. To the extent that you have only a small number of loci that theory would be violated and one would expect drastic changes in genetic variances and covariances.

TEMPLETON:

If you have a major gene system and a bunch of modifiers it turns out that very rapidly the only contributors to additive genetic variance are the minor genes. In fact you do not really deviate from the assumptions required of normal classical polygenic inheritance. You can estimate heritabilities from parent-offspring regression, etc. You see the effects of the minor genes, mostly. It is kind of a paradox—the "major" gene is not the major contributor to the additive genetic variance.

LANDE: How can it be a major gene if it is not a major contributor?

TEMPLETON:

You have to recognize the difference between the genotype-phenotype relationship and the breeding value. The major gene very rapidly gets equilibrated.

LANDE: So if it is at low frequency it remains a minor contributor to additive genetic variance even though it has major phenotypic effects.

TEMPLETON:

Even if the locus is in a polymorphic state, it rapidly gets to a kind of quasiequilibrium state and does not contribute much to variance in breeding values. The heritability of malarial resistance in sickle cell anemia at equilibrium is 0.

LANDE: But that is different from a major single gene influenceing some morphological trait. There would still be a major effect on the variance of that trait.

TEMPLETON:

Yes, it depends on the trait measured. If the traits relate to fitness the additive genetic variance would be almost exclusively from minor loci.

ROSE: You are talking about the approach to equilibrium under selection?

TEMPLETON:

Yes. You do not need exact equilibrium for most of the additive variance to be contributed from minor loci. In fact, you do not get a major deviation from the usual polygenic models in terms of measuring response to selection, for example.

C. A. TAUBER:

Ecologists should be alert to the kinds of characters they are measuring —traits like age at first reproduction. We need to determine what is controlling the trait. We need to know the physiological mechanisms behind the ecological characteristics before we can get to their genetic bases.

ISTOCK: I would like to follow that up. The choice of observation to make is critical when you decide which characters are most essential for study, as Dr. Tauber just emphasized. If we are to contribute something to the understanding of gene effects by our statistical treatments of observations we need to extend our methods to deal with major genes imbedded in a system of minor genes. Thoday worked on this several years ago and made many of the points being discussed here. The real question remains what characters to pick. Some of the traits ecologists love, like age at first reproduction, are natural ones to deal with and Steve Arnold said they are the worst from the quantitative genetics standpoint. Is it simply that you would rather work with morphology or is there another reason you said that Steve?

ARNOLD: I succeeded in being provocative, didn't I? We are, of course, interested in ecologically relevant traits because they are so important. They are, as Wright put it, the ultimate characters. We can honestly believe that all other characters funnel their effects through these major components of fitness. I was making the simple methodological point that these characters violate more of the classical assumptions of quantitative genetics than any we could invent! The observation is just that analyses of these traits will not be easy. It will be exciting and it is important.

ISTOCK: To me, it seems the other way around. As Dr. Lande has pointed out, the theory is based on large numbers of alleles creating deviations and since traits like survivorship are influenced by many loci they might be the nicest for analysis.

ARNOLD: In many ways they are!

ISTOCK: It's no fun winning arguments!

LANDE: I think Dr. Arnold was alluding to the fact that many characters of
            most interest to ecologists are characters which change with age and are
            expressed differently in different environments. This particularly true
            of behavioral traits. Other major components of fitness, like fertility
            schedules, are also subject to environmental effects and are also prob-
            ably subject to genotype-environment (GE) interactions. GE inter-
            actions are not directly addressed in quantitative genetics. They just
            end up as interaction sources of variance.

ISTOCK: On the other hand, we have tended to believe, on the basis of relatively
            limited evidence, that these are really messy characters. But as you have
            pointed out, they are seldom explored. How messy are they, really? If
            we set up a program to explore the structure of quantitative characters
            from an evolutionary perspective could we define environmental norms
            and use them in laboratory experiments to resolve these so-called messy
            characters?

ROSE: But, surely one of the most fundamental problems is that almost all
            natural environments oscillate to some degree, if only during the day. If
            you try to measure a quantitative character with a nice pattern of oscil-
            lation in the environment, you need a tricky method.

ISTOCK: Yes, that is one of the reasons I erected this static notion of the persis-
            tent environment rather than a constant environment. Ecologists get
            around messiness by continuing to be messy. Environments may fluctu-
            ate but continue to show stable averages and limited ranges.

ROSE: Here is one example of why that will not solve the problem. If you have
            periodicity in your environment, and periodicity in a character like egg
            laying, you create a mess if you try to handle your females sequentially
            because of the different periodicities.

ISTOCK: I agree. Parent-offspring methods really suffer from that problem but
            sibling designs may help resolve it.

ROSE: What I mean is you are still forced to measure females in succession
            through the day. You cannot handle them all at one time, instantane-
            ously.

SCHAAL: Implicit in a lot of what has been said here is the concept of a "trade-
            off." A great deal of reproduction may be achieved with sacrifice of a
            certain amount of survivorship. But maybe there is not always a trade-
            off. Why do you see trade-offs sometimes and not others?

GIESEL: One idea that I would not like to explore in public, is that *Drosophila* populations contain numbers of lethal, semilethal, and sublethal gene forms and their frequency varies among populations. Might you, in an unbiased, random collection of flies, expect to see positive genetic correlations—in the absence of environmental factors like severe food limitations—because there is a range (in normal populations) of genotypes from very poor to very good. Translation of genotype through physiology might lead to positive genetic correlations on the basis of something like an efficiency model. A fly with poor metabolic machinery might show low developmental rate, low fecundity, etc., just because of deleterious alleles.

STEPHEN HUBBELL (University of Iowa):

We are running in to one of the problems of evolutionary ecologists making cross species comparisons with no genetic experiments. They make the assumption that a negative trade-off occurs because all these species have the same energy budget and just partition the same energy, either into large numbers of seeds that are small or a small number of large seeds. The major genetic question is how the architecture within a species operates. The "trade-off" issue between species may be a myth.

PATRICIA WERNER (Michigan State University):

I want to underline that too! We are going to have to learn a great deal more about the particular organism we are dealing with. For instance, Dr. Schaal's plant versus my plant. I think if we have an indeterminant flower, we would not expect any kind of negative relationship between seed weight and seed number. If we had a determinant flower, we might expect such a relationship. Also, many of the positive correlations that are obtained in plant work are really related to size, including many of the life history relationships. So what you have to look at is what determines size. Very often those factors are simple gene changes that might influence, say, a branching pattern (or allocation or biochemistry). For example, *Solidago speciosa* seems to be different from all the other *Solidagos* just because of one difference in branching which gives it an entirely different form and severely limits height, and many other things. So, positive phenotypic correlations are not as important as the basic changes which give you negative correlations. Those are the things I am dealing with.

ANTONOVICS:

I think what you are getting at is a little bit of, as it were, naive quantitative genetic theory. The important trade-offs then come under a system of directional selection. If you are going to talk about genetic trade-offs, and this includes the additive genetic variance of the characters that are occurring, then it leads us to certain things. First of all, if you have a genetic polymorphism and the character is not under strong directional selection, then you would expect positive genetic correlations. Second, the way that negative correlations arise is that any genes contributing positively to fitness will be fixed. So, if you have none of the genes contributing negative effects, you will in fact wind up with fixation. Within-species genetic correlations may have no bearing on between-species relationships, because each of those species may have gone to fixation for different genes. We do not necessarily expect negative genetic correlations always. We would only expect it if we were dealing with a population at equilibrium and under strong stabilizing selection.

DOYLE: Dr. Antonovics has traced the trade-off concept to quantitative genetics, but it has always seemed to me that the concept, probably gratuitously, comes from ecological theory. Namely, the idea that you cannot simultaneously optimize everything. When a bird commits itself to a particular bill size, for instance, it cannot at the same time, take smaller or larger seeds.

ANTONOVICS:
            Yes, but I think you are dealing with a different situation in trade-offs within species as compared to those between species. There is a different process leading to heritabilities between populations as opposed to within populations.

Index